ISBN 978-1-331-51766-5
PIBN 10200768

This book is a reproduction of an important historical work. Forgotten Books uses
state-of-the-art technology to digitally reconstruct the work, preserving the original format
whilst repairing imperfections present in the aged copy. In rare cases, an imperfection in
the original, such as a blemish or missing page, may be replicated in our edition. We do,
however, repair the vast majority of imperfections successfully; any imperfections that
remain are intentionally left to preserve the state of such historical works.

1 MONTH OF
FREE
READING

at

www.ForgottenBooks.com

———◇———

English
Français
Deutsche
Italiano
Español
Português

www.forgottenbooks.com

Mythology Photography **Fiction**
Fishing Christianity **Art** Cooking
Essays Buddhism Freemasonry
Medicine **Biology** Music **Ancient
Egypt** Evolution Carpentry Physics
Dance Geology **Mathematics** Fitness
Shakespeare **Folklore** Yoga Marketing
Confidence Immortality Biographies
Poetry **Psychology** Witchcraft
Electronics Chemistry History **Law**
Accounting **Philosophy** Anthropology
Alchemy Drama Quantum Mechanics
Atheism Sexual Health **Ancient History**
Entrepreneurship Languages Sport
Paleontology Needlework Islam
Metaphysics Investment Archaeology
Parenting Statistics Criminology
Motivational

BRITISH BIRDS' EGGS:

A HANDBOOK

OF

BRITISH OÖLOGY.

BY

ARTHUR GARDINER BUTLER, F.L.S., F.Z.S., &c.

Illustrated, by the Author, in Chromo-lithography.

LONDON:

E. W. JANSON, 35, LITTLE RUSSELL STREET.

F.W.

205450

INTRODUCTION.

About the year 1852 I first fell a victim to a mania for collecting; nothing came amiss, from a cricket to a coin; but the ruling passion at that time found its highest satisfaction in the acquisition of butterflies and moths, with a dash of beetles thrown in by way of a relish. With the old fever still lingering in my veins, I visited the seaside in 1871; to my intense disgust, hardly a specimen of any description was to be obtained. Feeling the impossibility of dragging out an inactive existence for four weeks with no alleviation but that offered by the morning swim, I began to cast about for something to give zest to my rambles, when the idea of forming a collection of British Birds' Eggs came to me as a welcome relief. Without delay I set to work in search of nests, having no knowledge to guide me, beyond what I had picked up hap-hazard during my entomological excursions, which amounted to little more than an acquaintance with the notes and general appearance of some of our commoner birds, and the size and colouring of a few of their eggs. As might be expected, the whole of the eggs first collected were blown with two holes, and the discovery that I could expel the contents, by means of a blowpipe, from one hole, was made in total ignorance that this plan had long been in general practice.

The fact that my birdsnesting was begun without previous study was ultimately the cause of the production of the present work. Knowing practically nothing about the nests which I came across, I took eggs from every nest that was in the least degree unfamiliar either in form or materials, as well as all varieties either of colour or shape, though deposited in nests apparently identical. The natural result was that, when I returned home, I possessed so many varieties which it was impossible to name from the figures given by Hewitson and others, that it became necessary to tax the good nature of several experienced oologists in order to determine them. As the result of several years' collecting convinced me that no published work was at that time of much use in enabling the beginner to name his collection; as, moreover, I was now becoming familiarized with the nests and eggs of most of the commoner species, I asked the opinion of several of our leading ornithological writers as to the likelihood of an inexpensive work, with numerous illustrations of common varieties, being acceptable. These gentlemen not only gave me encouragement, but promised to assist me both with advice and the loan of specimens. Accordingly no sooner had the question of expense been met by my enterprizing and most patient publisher Mr. Janson, than I began the preparation of the plates.

To draw an egg correctly is no easy task, as I found to my cost. Not only must all the spots be drawn reversed and in perspective, but (in order to give rotundity to the figure) the egg must be correctly shaded. After spoiling the appearance of my first six or seven plates by carefully indicating all the shadows visible upon them, and, instead of a rounded surface, producing a resemblance to a dish, a mushroom, or a plum, it became evident that most of these malformed appearances were the result of shadows cast upon the eggs by window-sashes, box-lids, and other objects which intercepted the light. It therefore became necessary to fall back upon lessons in model drawing which I had received in early youth ; and, as might be expected, from that time the plates gradually improved in appearance.

The illustrations for this book were drawn entirely in the evenings, and chiefly in the colder months, when there were few outdoor attractions ; they consequently occupied me for about seven or eight years. Until they were completed no attempt was made to prepare the letterpress, except for the occasional jotting down of an interesting fact.

Nearly the whole of my collecting having been done in Kent, and for the most part inland, there was little opportunity either of collecting eggs of many of the larger birds, or learning anything by personal experience respecting their nidification ; it therefore became necessary to draw largely upon the experience of many excellent observers whose ornithological works were at my disposal.

In the preparation of this Handbook I have been under great obligations to Mr. H. Dresser and Dr. Mason, who have both lent me many rare eggs for illustration, no less than forty-three having been from time to time entrusted to my care by the former gentleman. Mr. E. Bidwell, Mr. H. Druce, and Mr. H. Seebohm have lent me specimens and the last-mentioned gentleman has further greatly assisted me by the presentation of a copy of his most instructive ' History of British Birds.' Not a few rare eggs have been given to me by the Hon. Walter de Rothschild, the Rev. W. Bree, of Coventry, and others, whilst Mr. J. E. Harting has rendered me much service by supplying numerous interesting notes on the various species.

When collecting, I have been greatly assisted by many kind friends in Kent, amongst whom I wish specially to

thank Mrs. Smeed and the members of her family, who
have shown me the most generous hospitality, giving me
full permission to ransack the gardens and grounds and
help in so doing; to Mr. W. Drake and his cousin Mr.
Salter for similar kindnesses and for collecting and
sending me not a few nests after my return to town; to my
sister-in-law Miss Alicia Tonge for the practical aid and
encouragement which she has rendered from the com-
mencement of my work; to Dr. John Grayling, whose
companionship I have enjoyed in some of my country
rambles, and who has obtained for me several eggs of our
larger birds. Many friends have also allowed me to search
their gardens, plantations, farms and woods, with true
English open-heartedness; although I have not space here
to thank them all individually, I am none the less grateful.

I cannot close this introduction without thanking one of
my London friends, Mr. Oliver Janson, who has at all
times been ready to help me, has obtained me many a good
nest and, in a word, has shown as much interest in my
work as my very excellent publisher.

The species are figured upon the plates nearly in the
same order as that adopted by Hewitson in his ' Illustrations,' but a few alterations have had to be introduced in
the letterpress in order to bring species of a family together
in accordance with the views of modern scientists. A
more correct classification is introduced, for guidance in
arranging the species in the cabinet.

A. G. BUTLER.

BRITISH BIRDS' EGGS.

Family FALCONIDÆ.

GOLDEN EAGLE.

Aquila chrysaetus, *Linn.*

Pl. I., fig. 1.

Geogr. distr.—Almost entire Palæarctic Region, extending southwards into temperate America : it is not uncommon in the highlands
of Scotland (where it breeds), but is rare in England and Ireland,
though it probably still breeds in the wilder parts of Mayo and Donegal.

Food.—Birds and small mammals, such as hares, foxes, lambs,
fawns, &c.

Nest.—Formed of branches, measuring from 5-6 feet in the greatest
diameter ; covered with sprigs of fir, heather, fern-tops, grass, moss,
&c. ; the central concavity shallow.

Position of nest.—Near the top of an almost inaccessible rock,
especially if well clothed with vegetation, or occasionally in a somewhat large tree

Number of eggs.—1-3 ; rarely 4, usually 2.

Time of nidification.—III-IV ; April.

I have only once had the pleasure of seeing this bird
at large, and that was some years ago when staying at
Chamounix with the late Mr. Hewitson ; the bird was
sailing round in a circle at an enormous height, and looked
no larger than a blue-bottle fly ; but every minute or so it
uttered its strange screaming cry, which sounded distinct
enough in that clear air. Hewitson says that Foula, one
of the Shetland Islands, is the favourite resort of the
Golden Eagle, and that nests formed in these islands are
constructed of the " rope-like pieces of sea-weed, which,
having their roots at the bottom of the sea, rise like mimic
forests to its surface, and spread out their long riband

B

leaves;" that they are, moreover, lined with roots, straw, dry grass, and wool.

The eggs, owing chiefly to the difficulty of obtaining them, are much valued by collectors, and command a tolerably high price in the market; they vary from a type blotched not unlike that of the Common Buzzard, as represented on my Plate, to a nearly pure white.

The legs of this species are feathered to the toes, and these have each only three scales. (See J. E. Harting, ' Zoologist,' 1867).

Eagles are remarkable for longevity and their capacity for long abstinence from food : an instance of one enduring hunger for three weeks is recorded by Pennant.

WHITE-TAILED OR SEA EAGLE.

HALIÆTUS ALBICILLA, Linn.

Pl. I., fig. 2.

Geogr. distr.—Entire Palæarctic Region; in Asia, southwards to India and China; also N. Africa; formerly local but not rare, and resident, in Great Britain, but now rarely seen in the summer.

Food.—Fish, birds, young animals, as lambs, fawns, &c., and sometimes carrion.

Nest.—A massive structure of sticks, lined with straws, plant-stalks, dried grass and moss.

Position of nest.—Generally near the coast, on a bare rock, or in lofty pine or other tree.

Number of eggs.—2.

Time of nidification.—IV-V.

The nest of one of these Eagles, discovered many years ago in Norway, by Mr. Hewitson, "was placed in a hollow of the rock, and was composed of a large mass of sticks, and appeared to be thickly lined with soft materials"; this nest contained an unhatched egg and a young eaglet.

The Sea Eagle used formerly to breed in not a few localities in the British Isles, such as Westmoreland, Cumberland, the Isle of Man, Dumfriesshire, East Galloway, on Ailsa Craig and the Bass Rock. It still breeds in the Highlands and the islands off the Scotch coast, but, as already stated, it is far more rare than formerly; the bird, however, is not uncommon on our coasts, especially in the autumn and winter months; it is mobbed as soon as it appears by flocks of Gulls, and when it ventures inland is similarly accompanied by Rooks.

The nest is said to be very similar to that of the Golden Eagle, but the eggs are of a pure white colour; that which I have figured is in the collection of Mr. Dresser, as also is the egg of the Golden Eagle upon the same Plate; an egg, said to be that of *H. albicilla*, in my collection, is of a more elongated shape than usual, and, as I do not know its history, it would not have been safe to figure it.

This bird differs from the preceding species in its bare tarsus, and in having eight scales on the first to third toe and nine on the middle toe.

OSPREY.
PANDION HALIÆTUS, *Linn.*
Pl. II., fig. 1.

Geogr. distr.—Europe, Asia and Africa generally; also N. and S. America, Australia and New Zealand; in Great Britain it breeds in Scotland.
Food.—Fish.
Nest.—Formed of stout sticks, piled up to a considerable height, mixed with turf, sea-weed, or heather and earth, the lining (which is flat) composed of moss and sometimes vegetable fibre.
Position of nest.—On trees or rocks, or the top of an old ruin.
Number of eggs.—3-4; generally 3.
Time of nidification.—IV-V. May.

If in well-wooded districts, the Osprey usually selects an aged tree for nesting purposes ; for fishing it usually confines its attentions almost exclusively to some one large piece of water until the supply grows scanty. I have seen what Hewitson believed to be this bird fishing at Lucerne.

According to Newman, *P. haliætus* probably breeds at Killarney and Loch Lomond; his description of the nest, however, seems loose, and that of the egg, " of an elliptical form, rather less than those of a Hen," incorrect ; it, however, formerly used to breed in Loch Lomond, Loch Awe and Killehurn Castle, and Loch Menteith ; at the present time it probably still breeds in Inverness and Rosshire.

The nest is in the form of a truncated cone, from which the sticks are said to project but little at the sides, " the summit is of moss, very flat and even, and the cavity occupies a comparatively small part of it " (Yarrell).

The following I quote from Seebohm's 'History of British Birds ' :—" Its plumage is unusually dense on the lower parts, as a protection against its repeated immersions in the water ; and the long feathers adorning the tibiæ of the land Raptores are in the Osprey replaced by short ones. From the peculiar form of its finny prey, the slippery nature of its outer surface, and its great facility of evading the bird's attack, the Osprey's feet exhibit certain well-marked peculiarities. The outer toe is reversible, the claws are remarkably curved and sharp, and the soles of the feet are very rough, all assisting the bird to grasp its food with great certainty and precision " (vol. i. pp. 58, 59).

KITE.

MILVUS ICTINUS, *Sav.*

Pl. II., fig. 2.

Geogr. distr.—Western Palæarctic region; wintering in S. Europe and N. Africa: formerly common in Great Britain, but now almost extinct.

Food.—Fish-refuse, reptiles, small birds and young of game birds and poultry, moles, mice, and rats.

Nest.—Similar to that of a Buzzard but larger; rather flat, formed of rough twigs and small boughs, and lined with grass, roots, tow, cloth, rags, paper, &c.

Position of nest.—On the forked branch of a tree at a considerable height from the ground.

Number of eggs.—3-5; rarely more than 3.

Time of nidification.—IV; end of April.

I once saw this bird circling over a poultry yard at Sittingbourne, in Kent, but probably it was not hungry, as it suddenly turned and soared away out of sight: when pressed by hunger the Kite has been known to be knocked down with a broom and killed, so intent was it upon its prey.

Owing chiefly to the cutting down of woods and forests, which has done more to rid our country of interesting birds than can be remedied by any Wild Birds Protection Act, and (in a less measure) to the persecution it has suffered at the hands of farmers and gamekeepers, the Kite has been almost exterminated as a habitual breeder; it is, however, supposed still to hold its own in the counties of Aberdeen, Perth, and Inverness.

" The Kite," says Seebohm, " may be easily distinguished upon the wing by its deeply-forked tail and the peculiar nature of its flight. For hours this bird will keep the air, sailing in slow circles without an apparent effort, with wings and tail expanded to their fullest extent, the latter ever and anon being twisted to assist it to describe its graceful curves. From this habit of sailing in the air the Kite has gained the almost universal name of ' Glead'—a corruption of the word ' glide,' which aptly represents its beautiful aërial movements." (Hist. Brit. Birds, vol. i. pp. 75, 76).

COMMON BUZZARD.

BUTEO VULGARIS, Leach.

Pl. II., fig. 3.

Geogr. distr.—Europe generally, not extending far into Asia; rarely visiting Africa; in Great Britain it breeds in the western and northern counties of England and in Scotland, but it is not common.

Food.—Insects, reptiles, small birds, including young poultry and game, small mammals (as moles, field mice, rabbits, and leverets).

Nest.—Of large sticks lined with wool, fur and other soft materials; or an old Crow's nest lined with feathers, grass, &c.

Position of nest.—In the forked branches of lofty trees, such as beech or oak-trees; on rocks or the edges of steep banks bounding a stream or torrent.

Number of eggs.—3-4.

Time of nidification.—III-V; end of March to beginning of May.

In England this species appears to be confined to the larger woods during the breeding season; it is a dull sleepy kind of bird, flying as a rule slowly and only for short distances, though sometimes it ascends spirally to a considerable height; as, however, it does not pursue its prey upon the wing it is more destructive to young game and poultry than to full-fledged birds, which it only preys upon when they are wounded and therefore helpless.

A singular instance is recorded by Mr. Cecil Smith (Zool. 1878, pp. 339-40) of the building of a nest by this bird apparently as a ruse to mislead those in search of its eggs: according to his account a nest had been built on the previous year from which the eggs had been taken; the same birds had built a second nest close by, and though they deposited no eggs therein they did their utmost to prevent its being examined: in the meanwhile, however, they were quietly incubating a clutch deposited in the old nest. Such a proceeding points to something rather higher than blind instinct, unless indeed the eggs were laid in the wrong nest owing to deficient memory on the part of the parents: I have on several occasions known Blackbirds to visit an old nest after building and furnishing a new one, and the behaviour of the Buzzards, though differing somewhat, may have been due to a like cause.

HOBBY.

Falco subbuteo, *Linn.*

Pl. III., fig. 1.

Geogr. distr.—Palæarctic Region generally: not common in the British Isles, owing to the incessant war waged with it by keepers and others.
Food.—Insects, small birds, pigeons, &c., and young of larger birds.
Nest.—An old Crow's, Magpie's, or Dove's nest repaired.
Position of nest.—In isolated groves of tall fir-trees, upon projecting ledges of rocks, the ridge of a hill, &c.; usually in open woods or more or less overhung with brush.
Number of eggs.—2-5; generally 4.
Time of nidification.—V-VI; June.

According to Hewitson this species is most common in Yorkshire;* it is a bold and beautiful bird, and though it undoubtedly does occasionally destroy a partridge, it usually flies at smaller game, such as sky larks or beetles, but most frequently the former.

The Hobby has been trained to fly at Larks, Quails, and Snipes, but it is said to be not strong enough to be efficient in the field; it arrives in this country in April, and generally leaves again in October.

In its appearance and habits the Hobby resembles the Peregrine Falcon, excepting that it is much smaller; its flight is extremely rapid, cutting the air like that of a Swallow, upon which bird it appears, from a paragraph in Mr. Seebohm's 'History of British Birds,' it occasionally pre s.

This species, now so rare in our country, has been met with in China, India, Persia, Palestine, the Blue Nile, the Cape, Morocco, and perhaps Algeria.

---

* In the revised edition of Yarrell's 'British Birds,' however, we read that "In Somerset, Mr. Cecil Smith says that it is a very rare bird. It does not seem to be much commoner in Dorset or Wilts; and thence Oxfordshire, Northamptonshire, and Lincolnshire seem to form the north-western frontier of the district in which it can be said usually to breed, though instances are known of its having done so in Nottinghamshire, Lincolnshire, and Yorkshire."

MERLIN.

Falco æsalon, *Tunst*.

Pl. III. fig. 2.

Geogr. distr.—Europe generally, wintering in N. Africa; in Asia as far east as China; breeds in N. Siberia: in Great Britain it is said to have bred in Hampshire, Essex, Herefordshire, Shropshire, Pembrokeshire, Derbyshire, N. Wales, and regularly in the north of Scotland and Ireland.

Food.—Small birds, mice, and shrews.

Nest.—Usually a mere depression in the ground, into which the surrounding grass and heather has been scratched as a lining; rarely (in trees) formed " of sticks thickly lined with wool."*

Position of nest.—Amongst tall ling upon heathy moors; or in forest trees.

Number of eggs.—4-6; usually 5.

Time of nidification.—V.

In Belgium this bird's favourite food appears to be Quails and Partridges; and the hen bird is said to pounce upon them in England, but I have had no opportunity of testing the point : one thing, however, is certain, its habit of skirting hedges in search of prey is much more likely to provide it with warblers and finches than with partridges; though, when skimming low over the open ground, it is quite as likely to obtain the latter.

The Merlin breeds regularly in all the northern counties of Great Britain from Yorkshire to the Shetland Isles, but the nest is only occasionally found in the southern counties : Mr. Seebohm says :—" The Merlin's haunt in the breeding-season is indeed a wild and lovely one, amongst the remotest parts of the moors, where the silence is rarely broken, save by the notes of those few birds who share its favourite solitudes—the Red Grouse, the Moor Pipit, the Curlew, and the Snipe. A true bird of the mountain indeed it is ; and the observer must therefore be prepared for a long tramp over the heather, and doubtless a wetting from the mists which so frequently enwrap its breeding-grounds, if he wishes for a sight of its beautiful eggs and scanty nest." (Hist. Brit. Birds, vol. i. p. 37).

---

* Hewitson.

KESTREL.

FALCO TINNUNCULUS, Linn.

Pl. III., figs. 3, 4 ; and Pl. XXXVII., fig. 3.

Geogr. distr.—Entire Palæarctic Region ; wintering in India and Africa : resident in Great Britain.

Food.—Insects, reptiles, small birds, mice, and shrews.

Nest.—Frequently the deserted nest of a Crow, Magpie, Jackdaw or Raven ; but frequently in holes : it is formed of sticks and lined with wool or other soft materials, as grass, moss, &c.

Position of nest.—In trees, or holes in trees, ruins, belfries, towers of churches, or under eaves of old buildings.

Number of eggs.—4-5.

Time of nidification.—V.

I have in my collection two eggs of this species which were laid by a bird formerly in the possession of my friend Dr. Grayling, of Sittingbourne, in Kent ; he writes to me respecting them (Oct. 30th, 1877) as follows :—

" I wish I could do justice to the memory of the two Kestrels : these birds, male and female, were brought to me, with three others, funny little nestlings ! For a time they became very domesticated, but occasionally wandered into neighbours' premises, requiring many shillings for their recovery ; eventually I let them go. For a few days they hovered over their old haunts, came to be fed, and finally took to the woods, and I hope are still in existence.

" The female laid several eggs, two of which you have : we taught the birds to sit on the finger (with gloves, mind you). Now and then, when not regularly fed, they would pounce down on a flock of Sparrows, but usually mice and small Linnets constituted their daily meal.

" Waterton places this bird in his true relation to the interests of the farmer, ' The Winhover is a mice-destroying Falcon.' "

Cases have been recorded of the Kestrel killing and devouring its mate, but such instances are probably not frequent. In ' The Zoologist,' vol. iii., p. 936, F. Holme states that five males were successively trapped from the same nest, proving that there is always a reserve of unpaired birds to supply deficiencies caused by death : the unfaithful partners probably had this fact under consideration when enjoying their cannibalistic feast.

HONEY BUZZARD.

PERNIS APIVORUS, *Linn.*

Pl. III., fig. 5.

Geogr. distr.—Western Palæarctic Region, migrating to Africa for the winter ; very rare in Great Britain.
Food.—Insects, reptiles, small birds and mammals, as rats, mice, young rabbits, &c.
Nest.—Formed of twigs lined with fresh leaves or wool, or the deserted nest of a Kite or Buzzard similarly lined.
Position of nest.—In trees (near to the trunk) at a height of from twenty-five to fifty feet from the ground in groves or dense forests.
Number of eggs.—2-4 ; generally 2.
Time of nidification.—V-VI.

It is supposed that this bird was formerly commoner in the British Islands than at present ; any way, it is now a rare bird with us : it has bred in Northumberland, Staffordshire, Shropshire and Northamptonshire in comparatively recent times, and it still breeds in the New Forest. A few instances of its breeding in Scotland and Ireland have been recorded.

It is said that the Honey Buzzard generally builds its nest in a tall oak, between twenty-five and fifty feet from the ground, and that it is composed of dead sticks (some of them as thick as one's finger), to which lichens are adhering ; that the interior is formed of twigs, and lined with wool and fresh oak and beech leaves ; also that the parent birds surround it with a kind of bower of leafy boughs, which are renewed as the leaves wither.

The name of Honey Buzzard was given to this species on account of its habit of destroying the nests of bees and wasps for the purpose of feeding upon the grubs contained in the cells.

This species nearly resembles the Common Buzzard, but, according to Seebohm, " may at once be distinguished by the scales on the tarsus, which are finely reticulated all round instead of being in broad plates at the front and back." The lores are also " finely feathered down to the cere instead of being only covered with bristles."

PEREGRINE FALCON.
FALCO PEREGRINUS, *Tunst.*
Pl. III., fig. 6.

Geogr. distr.—Greenland to Southern Africa; Asia as far as Java and Sumatra; America from Hudson's Bay to the Argentine Republic: it occurs in suitable localities in Great Britain, being found on the rocky coasts of Scotland, England from Kent to Cornwall, and Wales; less commonly in Ireland.

Food.—Small birds, young game and ducks.

Nest.—Loosely formed of twigs and sticks.

Position of nest.—In deserted nests of other Raptores, in trees, on inaccessible cliffs and ledges of rocks, or even occasionally upon the ground.

Number of eggs. - 3-4; usually 4.

Time of nidification.—III-VI; May.

This bird has been known to attack the Heron, being both powerful and courageous : it is therefore not surprising that it was formerly much valued by falconers : it breeds chiefly on the ledges of bold cliffs on most parts of our coasts, and consequently its eggs are not, as a rule, easy to obtain.

The speed of the Peregrine Falcon upon the wing has been estimated at probably not less than a hundred miles an hour, and the force with which it strikes its quarry something prodigious : unlike some of the other hawks, it kills its victim at once, by tearing open the throat or head. Like all the hawks, it is said "invariably to choose the easiest flight," and therefore, as the weaker game are destroyed before the stronger, it is believed that the constant destruction of hawks of all kinds by gamekeepers is positively injurious to the interests of preservers. Of course there are two sides to this question, for it cannot be denied that such an argument will only apply to full-fledged birds pursued upon the wing, and not to such as are taken from the nest or poultry-yard. According to Dixon, the Puffin is a favourite morsel, but nevertheless he says that this bird displays very little alarm whilst the Peregrine is sailing high in air above it, and moreover that it is not uncommon to see it, "when its appetite is satisfied, surrounded by Terns and Gulls, and see the Puffins sitting quite unconcerned a stone's throw from their enemy."

SPARROW HAWK.
ACCIPITER NISUS, Linn.
Pl. IV., fig. 1, and Pl. XXXVII., fig. 2.

Geogr. distr.—Entire Palæarctic Region, especially Europe : generally in Great Britain, but more commonly in wooded and cultivated localities.

Food.—Beetles, Mollusca, lizards, small birds, young game and poultry, moles, and mice.

Nest.—A flat and shallow structure similar to that of the Ring-Dove, but larger; composed of slender twigs : or a deserted crow's nest lined with moss and hair.

Position of nest.—Generally on lofty cliffs or trees, but sometimes on low trees, or even thorn bushes.

Number of eggs.—4-7 ; rarely more than 5.

Time of nidification.—IV.

In an article in the 'Zoologist' for 1878, p. 347, Mr. C. M. Prior says of this bird, " Almost all of the nests I found were undoubtedly of the Sparrow Hawk's own make, one or two were in Crows' nests, and I never found one in a Magpie's. They generally preferred fir tree, especially Scotch, and, unlike the Kestrel, they do not like building in an open country ; the latter appear indifferent, but Sparrow Hawks greatly prefer a fir tree spinney. The eggs vary in number from four to six ; the last nest I found was in a larch tree, and contained the latter number."

" The young seem to be fed in a large measure on Missel Thrushes ; and the latter bird, more than any other, serves to satisfy the hunger of the parents. They are also very fond of little rabbits, greenfinches, and sparrows."

The Sparrow Hawk is also very destructive to young poultry, flying low over a yard and seizing a chick almost before one is aware of an invader ; this bird, therefore, is greatly detested by the farmer, who loses no opportunity of destroying it.

The eggs vary considerably in the number and depth of colour of their markings; that figured on Plate XXXVII. is one of a clutch taken by Mr. Edwin Shepherd, and given to me by him. I also have eggs almost wholly white.

MARSH HARRIER.

CIRCUS ÆRUGINOSUS, *Linn.*

Pl. IV., fig. 2.

Geogr. distr.—Southern and Central Europe, ranging into Africa as far as the Transvaal, eastward as far as India; occurs also in Japan; formerly numerous in the few districts of Great Britain, but is now very local.

Food.—Worms, Mollusca, fish, reptiles, small birds and water-fowl, rats, rabbits, and carrion.

Nest.—A large, flat clumsy structure of dead flags, reeds and sticks, the central cavity very shallow.

Position of nest.—On the ground or amongst dense masses of partly submerged flags or reeds.

Number of eggs.—4-5; rarely 6.

Time of nidification.—V.

This Falcon is said occasionally to breed in the fork of a large tree, in which case the nest is formed of sticks: it frequents desolate moors, marshy land, and barren, sandy wastes; its flight is not very rapid, and usually low, though in the breeding season the male is said to soar to a considerable height, and remain suspended on the wing for a great length of time while the female is sitting.

Speaking of the nest and eggs, Hewitson says, "In the fen countries, its usual resort, the nest is composed of so large a quantity of flags, reeds, and sedges, as to raise it a foot or a foot and a half above the ground. The eggs are usually four, sometimes—though not often—five in number, the time of incubation early in May."

The only regular breeding-places of the Marsh Harrier in England are now said to be Devonshire and the eastern part of Norfolk; and, in Scotland, Aberdeenshire; it however occasionally breeds in other counties.

The eggs of this and the other species of Harrier upon my plate were lent me by Mr. Dresser; I have, therefore, no doubt of their authenticity; I have eggs in my own collection, but none of my own taking.

HEN HARRIER.

CIRCUS CYANEUS, L*inn.*

Pl. IV., fig. 3.

Geogr. distr.—Europe generally, ranging southwards to N. Africa and eastwards to China and Japan ; not rare in Great Britain, breeding chiefly in the northern counties ; but rare in England.

Food.—Reptiles, small birds, rats.

Nest.—A few twigs or heather-stalks and a little coarse dried grass; but when in a damp situation it is placed upon a mass of flags, reeds, &c., which raise it considerably above the surface of the ground.

Position of nest.—Always upon or near the ground, and frequently in marshy situations, but usually on a dry moor.

Number of eggs.—4-5 ; generally 4.'

Time of nidification.—V-VI.

Though a rare bird in this country the Hen Harrier is still believed to breed in Devonshire, Somersetshire, Dorsetshire, Hampshire, Sussex, Kent, Norfolk, Gloucestershire, N. and S. Wales, Shropshire, Yorkshire, Durham, Cumberland, Northumberland, Wigton, Selkirk, Lanark, Haddington, Stirling, the Highland shires, Hebrides, Orkneys and Shetlands. Its rarity is believed to be chiefly " due to agricultural improvements which have brought into cultivation large tracts of what was formerly waste land " (Yarrell).

The flight of the Hen Harrier is said to be easy and buoyant, but not rapid, and, except in the breeding season, generally within a few feet of the surface of the ground, which it carefully examines in search of food. Mr. Seebohm says that it " is a bolder bird in the pursuit of its food than the other two British Harriers, and undoubtedly often chases its prey on the wing. It catches small birds, mice, frogs, but does not disdain to make a meal off the eggs of its neighbours when it has the opportunity. The gracefulness of its flight, and the ease with which it can skim over the brow of a hill, make it a favourite with the ornithologist, in spite of an occasional young grouse that may fall a victim to its prowess." (Hist. Brit. Birds, vol. 1. p. 130.)

MONTAGU'S HARRIER.

CIRCUS CINERACEUS, *Mont.*

Pl. IV., fig. 4.

Geogr. distr.—Europe generally, with the exception of the more northern countries; in Africa as far south as the Cape and the Comoro Islands, and in Asia as far east as China; in Great Britain it is less numerous than formerly, and local; if it breeds in Scotland it must be very rare.

Food.—Frogs, reptiles, small birds, water rats, and sometimes young hares and partridges.

Nest.—A mere hollow in the ground, with an outer border of heather-twigs and a lining of dry grass and sedge laid loosely together.

Position of nest.—In clearings amongst furze.

Number of eggs.—4-5; rarely 6.

Time of nidification.—V-VI; May,

The nest of this species is much less perfect than that of the two other species, and, in fact, hardly deserves the name; as with several other birds of prey, incubation frequently begins as soon as the first egg is laid.

It is believed that this bird still breeds sparingly in Pembrokeshire, Somerset, Dorset, Kent and Norfolk, but that there is no county in Great Britain in which it can now be said to breed regularly; in some counties where it used to abound it has now become an irregular autumn migrant.

Mr. Seebohm thus describes the discovery of a nest of this bird in a rye-field near Germany:—"The rye stood more than five feet high in a field which could not be much less than a hundred acres in extent. It seemed like looking for a needle in a haystack; but our host and guide told us that several pairs of Harriers bred annually on his farm; so we walked down each side of the rye, one of us following a narrow path up the centre. We saw at different times five or six birds, one pair especially seeming to show some anxiety at our presence. Finally one of the birds dropped somewhat suddenly into the waving corn; Dr. Blasius undertook to stalk her up, but, when she rose, missed her with both barrels. We were, however, delighted to find that she had risen from her nest containing four fresh eggs. There was no hole whatever in the ground; the rye had only been trampled down, and a slight but somewhat neat nest made of corn stalks, lined with a little dry straw. The nest was rather more than nine inches in diameter, and about two inches and a half deep in the middle." (Hist. Brit. Birds, vol. 1, p. 132.)

FAMILY STRIGIDÆ.

BARN OWL.

STRIX FLAMMEA *Linn.*

Pl. V., figs. 1, 2.

Geogr. distr.—Widely throughout Europe, India to Burmah, S. Africa; southward to Australia and N. and S. America; common and resident in England, Ireland, and Wales; less common towards the north of Scotland.

Food.—Insects, mice, shrews, rats, bats, and small birds.

Nest.—A mere hole.

Position of nest.—In holes in trees, deserted dove-cotes, upon walls under eaves, in barns, old ruins, towers, and steeples of churches.

Number of eggs.—5-6.

Time of nidification.—IV-XI; end of April and May.

This bird rears several broods in the year; it has been known to nest and rear its young in confinement; it feeds chiefly on mice, and therefore is one of the farmer's best friends.

Speaking of the long intervals which occur in the laying of a clutch of eggs of this and the other species of Owls, Hewitson says, " The Owls do not deposit their eggs as other birds for the most part do, in regular daily succession." I am satisfied, however, that this statement requires considerable modification, and that the regularity of the deposit varies with the weather; thus in cold, wet weather I have found, in not a few instances, in the case of buntings, finches, and other usually regular birds, that only two eggs have been deposited in a week.*

The eggs of Owls are, as a rule, more orbicular than those of other birds, but I have seen several of those of the Barn Owl similar in shape to that of my second figure; the latter was drawn from one of a clutch, all of the same elongated form, and was given to me by a friend who took it from the nest.

As the Barn Owl is not given to hooting, but only to screaming, it has earned the name of the Screech Owl; when in confinement it cannot bear the eye of man, but becomes restive when stared at, turns away, and if persistently watched hops about and screams irritably.

---

* Hewitson himself quotes an instance of the nest of the Barn Owl being found with six eggs (the full clutch).

FAMILY BUBONIDÆ.

TAWNY OR WOOD OWL.

SYRNIUM ALUCO, *Linn.*

Pl. V., fig. 3.

Geogr. distr.—Throughout Europe, northward to about 70° N. lat., southward to N. Africa, eastward as far as the Ural Mountains; occurs in all the counties of England, and probably Wales; rarer in Scotland, and very rare in Ireland.

Food.—Frogs, slow-worms, small fish and birds, bats, mice, rats, moles, and rarely leverets or rabbits.

Nest.—A mere hole in a tree, wall, or the ground; or occasionally the deserted nest of a Rook or Crow.

Position of nest.—In a hollow tree, especially when thickly covered with ivy, or in an old ivy bush; in old ruins, chimneys, barns and rabbit burrows; it prefers groves or woods of oak or beech trees.

Number of eggs.—3-5; rarely more than 4.

Time of nidification.—III-VIII.

Eggs of this species in my collection were taken in June in the New Forest, and were forwarded to me unblown. The Tawny Owl has, however, been known to lay as late as August.

Cases have been recorded of this Owl nesting in the deserted burrows of rabbits or foxes at about two feet from the entrance to the hole.

The Tawny Owl is the favourite of the poets, because it makes a blood-curdling hooting noise at night, which Hewitson says is very pleasant to the ears of "one fond of Nature's sylvan sounds." My old friend was particularly fond of the bagpipes, and I have often seen his look of pleasure when visited by an old Highlander armed with those elements of discord. I never heard the precise sounds attributed to the Tawny Owl, to issue from its throat; but they are about as near as most of the vocabulary attributed to animated Nature.*

The habit this Owl has of building in ivy has given rise to the well-known saying "You look like an owl in an ivy bush." Eggs have, however, been known to be deposited on the ground beneath the drooping branches of a fir tree.

Amongst the Mammalia one of the most notable is the "*Arrr-cke*" of the Ass, which has been rendered "*Hee-haw.*"

LONG-EARED OWL.

Asio otus, *Linn.*

Pl. V., fig. 5.

Geogr. distr.—Has a wide range over Europe, and visits N. Africa in winter ; is found in Asia as far east as China and Japan, and as far south as N. W. India. Resident in Great Britain, especially wooded districts in the North of England.

Food.—Insects, small birds, mice, rats, shrews, moles, young rabbits.

Nest.—The deserted nests of members of the Crow family or Buzzard, slightly repaired, and lined with feathers or down, or in Squirrels' nests.

Position of nest.—In trees in woods.
Number of eggs.—4-5 ; usually 4.
Time of nidification.—III-IV.

Speaking of the note of this species the Rev. W. W. Flemyng says (Zool. vii., p. 430), " I heard it several times utter, at intervals of about a quarter of a minute, a very plaintive single note in a high key, which I syllabled " moo." .... With regard to the very strange quacking note of the Long-eared Owl (described by Mr. Ussher, Zool. 1882, p. 265), is it not singular that this sound has not (apparently) been heard by English ornithologists, and that, as far as I am aware, no mention is made of it in any of the text-books on British birds ?"

Speaking of nests taken by the Messrs. Tuke, Hewitson says, " These nests were either on the outskirts or in an opening of the wood, the ground beneath them being strewed with the remains of the Blackbird, Yellow-hammer, Green Linnet, and Chaffinch."* It is probable that all these birds were seized when at roost, the Long-eared Owl being especially a " bird of night."

Asio otus inhabits Great Britain from Cornwall to Caithness ; it also breeds in some of the Hebrides, and has occurred in the Orkneys and Shetland : a letter from Mr. Isaac Clark, quoted in Seebohm's History of British Birds, states that it never builds a nest for itself, but always repairs an Old Wood Pigeon's, or Magpie's nest, and that the earliest date at which he had found a nest with young was on the 1st April.

---

* A correspondent of the late Mr. Gould also recognised the remains of the Wheatear, Willow Wren, and Bullfinch, and Mr. Newton traced the Goldfinch in the stomach of one which he examined. (See the 4th edition of Yarrell's British Birds).

SHORT-EARED OWL.
ASIO ACCIPITRINUS, *Pall.*
Pl. V., fig. 4.

Geogr. distr.—Europe, Asia and America generally: from the Arctic regions to Africa and lower southern America; also the Hawaiian Islands, but not Australia; in Great Britain, though widely distributed, it breeds sparingly, more commonly in Scotland than in England, where it is chiefly confined to the northern counties.

Food.—Mice, shrews, rats, moles, young rabbits, and young birds (but chiefly field mice and rats).

Nest.—A mere depression in the ground, sometimes lined with a little loose dry grass.

Position of nest.—Amongst grass, heather, or reeds, on open moors, in stubble, grass, or turnip fields.

Number of eggs.—4-7.

Time of nidificaton.—IV-VI; May.

Though this bird is only known to breed upon the moors in the north of England it occurs in the southern counties later in the year, when it does good service in the destruction of field mice; at this time it is said that it is frequently seen by sportsmen when Partridge shooting, mounting high when disturbed, and apparently suffering no inconvenience from the daylight.

Describing the discovery of the nest of this species in Hickling Marsh, Mr. Seebohm says, "The eggs were extremely conspicuous from one side of the heap of rushes when the bird was off the nest; but so long as she sat close it might very easily be passed by without notice. Both birds were quite silent the whole time. . . . There was not much attempt at a nest, not more than the Peewit makes. The ground seemed to be trodden into a hollow, which was lined with a few dry broken reeds and sedges. The reeds were lying in a heap on the ground; and in the place selected for the nest the thick cut ends slightly overhung the base of the heap, and formed some shelter over the nest on one side. We found a second nest on the following day containing six fresh eggs. It was in a part of the marsh where there were very few reeds, the ground being covered with *Carex* and *Juncus*. This nest was very similar to the one we found on the previous day, and was lined with flat leaves of *Carex*, with a feather or two, and was surrounded with a few slender willow bushes." (Hist. Brit. Birds, vol. i. p. 171.)

FAMILY **LANIIDÆ**.

RED-BACKED SHRIKE.

LANIUS OOLLURIO, *Linn.*

Pl. VI., figs. 1-6 ; and Pl. XXXVII., fig. 7.

Geogr. distr.—Europe in the summer, as far east as the Ural Mountains, arriving in April or May and leaving again in August ; N. East and South Africa; generally distributed in Great Britain.

Food.—Insects, small birds, and mice.

Nest.—A rather large cup-shaped structure, not unlike some nests of the Greenfinch, but deeper ; formed externally of dry grass-straws and moss ; lined with fine bents, wool, and horsehair.

Position of nest.—Most frequently in a hawthorn bush or hedge ; sometimes in the fork of a dwarfed tree, but seldom more than five feet from the ground, and frequently less.

Number of eggs.—5-6; usually 5.

Time of nidification.—V.

I have found the nest of this bird tolerably common in Kent, more especially in the neighbourhood of Maidstone ; the nest, when placed in a tree, was invariably more solidly built than when in a hedge or bush ; when the full compliment of eggs has been laid, and the hen has begun to sit, she will show the greatest rage at the removal of her eggs.

The title of " Butcher-bird " has been earned by this bird, from its custom of spitting its prey upon thorns when devouring it ; curiously enough, though I have repeatedly found the nest, I never yet came across a single instance in which the larder was not empty ; the fact, however, is well attested that the Butcher-bird has acquired this savage habit in order the more easily to tear its prey to pieces.

The red variety of the egg appears to be by no means common ; I have only once taken a clutch, and on that occasion I had to fight my way through a very dense and tall hawthorn hedge in order to reach them, whilst, to add to my discomfort, I had to listen to the most unbridled language from the mother-bird.

The Red-backed Shrike is single-brooded.

FAMILY MUSCICAPIDÆ.

SPOTTED FLYCATCHER.

MUSCICAPA GRISOLA, *Linn.*

Pl. VI., figs. 7-10; and Pl. XXXVII., fig. 8.

Geogr. distr.—Europe generally as far as about 70° N. lat.; in Asia as far eastward as Dauria, and southward in Africa as far as the Cape; it arrives in Great Britain in April or May, and leaves again in August or September.

Food.—Insects.

Nest.—Composed of twigs and roots, or fine grasses, mixed with a quantity of green moss, interwoven with spiders' web; lined with fine grass, hair, and sometimes two or three feathers; varying in shape from a semicircle to a complete cup according to its position.

Position of nest.—Upon stumps of branches of old fruit trees, on the shattered and partly hollowed trunks of small trees, on branches of wall fruit trees (especially plum), in high hawthorn hedges, on low branches, or amongst roots of trees overhanging water, in holes in walls, in metal gutters on roofs of houses, &c.

Number of eggs.—4-5.

Time of nidification.—V-VI; June.

I have taken the eggs of this species as early as the 30th May, but it is rare to obtain them before June.

On the 4th June, 1878, I removed three eggs from a rather small nest of the Spotted Flycatcher formed in the hollow top of a tree-stump in a small plantation. I substituted for the eggs three hazel nuts, which completely filled the cavity of the nest. I returned on the 8th of June and found the bird sitting; she had ejected one of the hazel nuts to make room for her fourth egg. I briefly noted this fact in the 'Zoologist' for September, 1878, as an evidence of the fact that birds are unable to recognise their own eggs.

A large nest, chiefly lined with reddish hair, and containing two eggs as figured on Plate XXXVII., was taken out of a tall hawthorn hedge; I have described this nest in full in the 'Zoologist' for December, 1883.

A nest in my collection, of the shape of a slipper, was found in a hole in a wall in June, 1885, by my friend Mr. William Drake, of Kemsley, in Kent; the nest contains three eggs, which completely fill the cavity.

PIED FLYCATCHER.

MUSCICAPA ATRICAPILLA, *Linn.*

Pl. VI., fig. 11.

Geogr. distr.—Europe in summer; eastward as far as Persia; Africa as far southward as the Gambia in winter; in Great Britain it is local, especially in Scotland and Wales.

Food.—Insects.

Nest.—Loosely constructed of moss, rootlets, and dried bents, lined with wool, feathers, or hair.

Position of nest.—Usually in the deserted hole of a Woodpecker or Tit in an old oak, beech, aspen, or chestnut tree in a grove; rarely in a dense wood.

Number of eggs.—4-6; rarely as many as 8.

Time of nidification.—V-VI.

This species is said to return constantly to the same place to breed year after year, and, though usually a hole in a tree is selected for this purpose, the nest is sometimes placed in a hole in a wall or bridge, especially near to water; formerly it was supposed to be strictly a northern species during the breeding season, Cumberland and Westmoreland being its favourite places of resort, and Derbyshire being regarded as an improbable county in which to find the nest; in the 'Zoologist' for 1877, however, a specimen is said to have been seen in Wiltshire in April, and a second shot in Hampshire in May.* It is known to breed regularly in some parts of Yorkshire, Durham, in a few places in N. Wales and the English counties of the Welsh border, and occasionally in N. Devon, Somerset, Gloucester, Oxford, Dorset, the Isle of Wight, Surrey, and Norfolk; Mr. Hargitt has obtained eggs from Invernesshire, and it is said to be tolerably common in the Orkneys.

---

* The discovery of a nest would certainly have been more interesting, as the birds may have been passing northwards when seen; it has, however, been known to nest in Wiltshire.

FAMILY ORIOLIDÆ.

GOLDEN ORIOLE.

ORIOLUS GALBULA, *Linn.*

Pl. VI., figs. 13, 14.

Geogr. distr.—Central and South Europe, rare in West Europe; as far eastward as Turkestan; also in South Africa; rare in Great Britain: has bred on several occasions in England.

Food.—Insects in all stages, berries and fruit, especially cherries.

Nest.—Basket-shaped, firmly constructed, being woven on each side to a branch; formed of strips of bark, straws, dried grass, bents, &c.; ornamented externally with strips of white birch bark, and lined with fine grass bents.

Position of nest.—Suspended from one of the smaller branches near the top of a small tree in a dense wood or grove.

Number of eggs,—4-5.

Time of nidification.—V-VI; end of May and beginning of June.

Some years since (1868) I had the pleasure of seeing this handsome bird in the neighbourhood of Linton in Devonshire: it flew rapidly past, and the effect of the sun shining full upon its brilliant plumage was to me quite startling.

This species has been recorded as having bred in an ash plantation in Kent; the nest was of fibrous roots, and attached to two upright stems of ash; the structure was so thin that the young birds were visible through the bottom of the nest.*

In the spring the Golden Oriole has a soft warbling note, uttered between its ordinary cry, which is loud and shrill; the female is very attentive to its young, in defence of which, like the Blackbird, it shows great intrepidity: though a rare bird in this country, it is by no means uncommon in woods and gardens in France and Belgium, arriving in April and leaving in September.

Its conspicuous colouring probably does more to hinder its becoming common in this country than anything else, as it makes its appearance almost every spring in the northern and eastern counties, and in the west of England by forty at a time.

---

* This is, however, by no means the only recorded instance of its nesting in that county.

Family TURDIDÆ.

Sub-family CINCLINÆ.

COMMON DIPPER (or WATER OUZEL).

CINCLUS AQUATICUS, *Bechst.*

Pl. VI., fig. 12.

Geogr. distr.—Found in Asia Minor, Palestine, Persia, Yarkand, Dauria, Sikkim, and Algeria, throughout Central Europe, and in Great Britain; but local, owing to its affecting mountain streams.

Food.—Water insects, larvæ and pupæ of Neuropterous insects, such as caddis- and dragon-flies, fresh-water Mollusca, small fish, &c.

Nest.—Large, constructed of moss felted together and lined with grass stalks and dried oak or beech leaves; entrance in front.

Position of nest.—Carefully concealed in a hollow, under shelter of some overhanging rock or bank, frequently behind a water-fall.

Number of eggs.—Usually 4; rarely 5-6.

Time of nidification.—IV-V.

This species begins to build about the middle of April; it is said never to be seen in flocks, and rarely at any great distance from water; its song is a continuous subdued warbling, and it commences to sing as early as January; in its actions the dipper is quick and graceful, as it is beautiful in form and colouring; though I have only seen it in shallow streams, in which it ran about in search of insects, it is often found where the water is deep, and, when suddenly disturbed, is said to dive, using its wings to keep it below the surface.

The nest of the Dipper is described as like that of the Wren; to my mind this is rather vague, as two Wren's nests are rarely alike, and frequently totally dissimilar; a nest of the Great Tit in my collection, though a little smaller than that of the Water Ouzel, seems to me to be far more like it than most Wren's nests both in form and substance.

The Dipper has a smooth even flight, somewhat resembling that of the Kingfisher. Mr. More states that it breeds occasionally in Cornwall and Dorset, but regularly in Devon, Somerset, probably throughout Wales, Monmouthshire, Herefordshire, Shropshire, Staffordshire, Cheshire, Derbyshire, Yorkshire, Lancashire, Scotland generally, and the Hebrides; also suitable localities in Ireland.

SUB-FAMILY *TURDINÆ*.
MISSEL THRUSH.
TURDUS VISCIVORUS, *Linn.*
Pl. VI., figs. 15-17.

Geogr. distr.—Nearly all over Europe, ranges into the Himalayas (where, however, it has been regarded as a distinct species); breeds commonly in the British Islands, being found everywhere throughout England.

Food.—Berries (including small fruits), seeds, snails, worms, larvæ, and insects.

Nest.—Not unlike that of the Blackbird, large and heavy in con-struction; it is formed of twigs, roots, straws and grasses, with an inner lining of mud formed into pellets and mixed with grass or roots; the interior more carefully constructed of finer grasses and sometimes roots and moss; mosses and lichens are also occasionally attached to the exterior.

Position of nest.—In forks or on boughs of trees, generally not very far from the ground, in parks, orchards, plantations, and woods.

Number of eggs.—3-5; usually 4.

Time of nidification.—III-V; middle of April; rarely early in May.

I remember on one occasion seeing the nest of this bird in a fork near the top of a tall elm tree in Hyde Park; probably the bird had found by experience that, in such a public place it was of little use to build within reach of boys; a favourite position for the nest is in the hollow formed by the branching off of the boughs of an old apple tree.

The Missel Thrush has been known to lay twice in the same nest, which, however, was subsequently deserted owing to its cracking; it has sufficient affection for its young to induce it to attack larger birds which approach them too closely; the eggs vary a good deal in size and colouring, some eggs in my collection being marked not unlike those of the Blackbird, but having a more opaque appearance; the nest, however, appears to vary less, all that I have seen being heavily lined with mud, whereas this is not always the case with that of the Blackbird.

Mr. Harting has kindly called my attention to the fact that Mr. E. T. Booth once found the nest of *Turdus viscivorus* built in a small stunted bush within three feet of the ground. He had never previously seen one at so slight an elevation, and was of opinion that this lowly site was chosen as being less exposed to the attacks of Crows, which were very numerous in the neighbourhood.

SONG THRUSH.

TURDUS MUSICUS, *Linn.*

Pl. VII., figs. 1-9; and Pl. XXXVII., fig. 1.

Geogr. distr.—Found throughout the Palæarctic Region, but rare in the extreme east; migratory as a rule in Western Europe, though resident in some countries; breeds abundantly in the British Isles, where it is partially resident.

Food.—Beetles, larvæ, worms, snails, berries, seeds.

Nest.—A rather deep cup, very strong and heavy in construction; formed externally of slender twigs, roots, grasses, leaves, and moss, lined internally with a layer of mud, cow-dung, or rotten wood carefully smoothed so as to remind one of the interior of a cocoa-nut.

Position of nest.—In hedges, forks of young trees, ivy-covered walls, in crevices or under ledges of rock, among stunted willows near streams, òr at the root of a tuft of heath.

Number of eggs.—4-6; usually 5.

Time of nidification.—III-VIII; May and June.

Unlike the Blackbird, this bird usually goes very quietly off its nest when flushed; so far as my experience goes, it is less timid, and will sometimes allow itself even to be touched before deserting its eggs, possibly the conspicuous colouring of the latter may have something to do with this unwillingness on the part of the parent to expose them to view.

The nest is one of the commonest and most conspicuous objects which the collector can meet with; I have myself met with about thirty in a morning's ramble, and I have heard of as many as sixty being found in the course of a day: the wholesale grubbing of woods by agriculturists, which has, in my opinion, been the principal cause of the decrease in the numbers of many of our British Birds in late years, has apparently not in any way affected the Song Thrush, which, owing to its indifference as to the selection of a place for its nest, appears to become more abundant every year.

BLACKBIRD.
Turdus merula, *Linn.*

Pl. VII., figs. 10-18 ; and Pl. XXXVII., fig. 4.

Geogr. distr. Common and generally distributed nearly all over Europe ; partially resident in the British Isles ; it occurs in Palestine, Persia, Cashmere and Afghanistan.

Food.—Insects, spiders, grubs, worms, fruit, and seeds.

Nest.—Bulky ; round, semicircular, or oval, according to where it is placed, formed externally of stalks of grass and twigs twisted together and compacted with moss, usually with an inner lining of mud in pellets, roots, dead leaves, slender grasses, &c. ; in the neighbourhood of houses the outside is sometimes disfigured by the interweaving of old rags and pieces of newspaper with the twigs and moss.

Position of nest.—In hedges, shrubs, trees (especially trained fruit trees in orchards), holes in walls or rocks, or in low banks, where, however, it is less abundantly found.

Number of eggs. 4-6 ; generally 5.

Time of nidification. III-VII ; May and June.

Probably no nest is more easy to find than that of the Blackbird ; it is not only so abundant that I believe I have met with at least forty in a morning, but it is large, and often placed in the most conspicuous position. Though the parent birds are most devoted to their young, they appear not to have much affection for their eggs, for I have known the cock bird to devour them as fast as they were deposited, though, curiously enough, it only break-fasted off one each morning ; probably no bird has a poorer appreciation of form or colour, for the most irregular flint stones may be substituted for its eggs without causing it the least uneasiness—a fact well known to the rustics, who constantly take an egg out of the nest, and, sub-stituting the nearest stone, revisit it confidently on the following day, knowing that another egg will be ready to hand. The strange thing to me is that the discomfort of sitting upon a rough or sharp-edged substance fails to open the bird's eyes to the true state of the case. The white egg on Pl. XXXVII. was one of a pair taken from an ordinary nest at Wateringbury, near Maidstone, and presented to me by its discoverer.

RING OUZEL.

TURDUS TORQUATUS, Linn.

Pl. VIII., figs. 1-3.

Geogr. distr.—Asia and Africa; rare in Persia and Algeria; arrives in the British Isles in April, being most abundant in the more wild and mountainous districts; it has, however, been met with in Kent, Suffolk, Norfolk, Warwick, Leicester, and in the Isle of Wight.

Food.—Insects, Mollusca, fruits, berries.

Nest.—Much like that of the Blackbird, but somewhat looser in construction; formed externally of dry grass and bents compacted with a little earth, leaves, and moss; twigs of heather or larch are not unfrequently interwoven; the lining is formed of fine dry stalks of moor grass, and sometimes a little clay.

Position of nest.—Generally upon the ground, under a low bush, or in ling growing on the brink of an embankment or slope.

Number of eggs.—4-6; usually 4.

Time of nidification.—IV-VII; May.

The song of the Ring Ouzel is louder and harsher than that of the Blackbird. Though commoner in the northern than southern counties, I have taken it in Kent, both under a furze bush and from the edge of a heathery moor under overhanging ling at the top of a steep bank bounding a little frequented road; there is no mistaking the bird on account of the conspicuous crescentic white patch on its breast; this marking is, however, much less distinct or wanting in young birds. When disturbed from the nest the Ring Ouzel is as noisy as the Blackbird, and when the young are hatched it is equally bold in defending them; its eggs are like some varieties of those of the latter species, especially those which approach the Missel Thrush, and, occasionally, are hardly distinguishable from eggs of *T. viscivorus.* It breeds regularly in rocky parts of Cornwall, Devon, Somerset, Gloucestershire, Monmouthshire, Wales, Herefordshire, Shropshire, Staffordshire, Derbyshire, and thence northward to Caithness; in Ireland it is generally distributed in suitable localities during the summer months.

SUB-FAMILY *SAXICOLINÆ*.

BLACK REDSTART.

RUTICILLA TITYS, *Scop*.

Pl. VIII., fig. 11.

Geogr. distr.—South and Central Europe, eastward to Persia and southward to N. Africa: in Great Britain it is an occasional visitor, most frequent in autumn and winter; it is certainly rare in Scotland; in England it is said to have bred in Exeter, Staffordshire, and Notting.hamshire, and I have the egg taken in Hertfordshire.
Food.—Worms, insects, berries.
Nest.—Tolerably neatly constructed of dried grasses mixed with rootlets or moss, and lined with hair, wool, and a few feathers.
Position of nest.—In holes in cliffs, old walls or buildings near human habitations, or in a hole in a tree; also in hedges.
Number of eggs.—4-5.
Time of nidification.—IV-VI; May.

An egg of this species was exhibited at a Meeting of the Zoological Society in 1878 * by the Rev. R. P. Barron, M.A., as having been found by himself two years previously in a hole in an elm tree in his neighbourhood, and was recognised by Mr. Dresser and others as that of the Black Redstart. I had some conversation with Mr. Barron respecting this egg, and asked him to let me figure it in the present work, and he shortly afterwards sent it to me with the nest and the following notes :—

" The nest, I fear, is not very perfect, having been two years left in its place; it was found in the middle of May, 1876, right inside the hollow trunk of a living elm tree, at a distance of about seven or eight feet from the ground, on a projecting ledge of the inside wood, and within a few feet of a small lake. There were originally three eggs, of a slightly pinkish tint before being blown; they had been forsaken; the nest seemed to be lined with hair and hay. You need not, of course, return the egg or nest."

The nest appears to me to have lost much of its original bulk, the lining having probably been utilised by other birds before it was taken; the egg sent to me was of a bluish-white colour; it is figured on Pl. VIII., but the lithographic printing has exaggerated the blue tint which has now almost, if not quite, disappeared from the egg in my cabinet.

---

* Curiously enough this exhibition does not appear to have been recorded in the Proceedings.

In his 'Birds of Sherwood Forest,' pp. 67, 68, Mr.
W. J. Stirling records the taking of nests of this species
as follows :—
"My first acquaintance with it was the discovery, on
May 17th, 1854, of a nest in a thorn hedge by the side
of the road leading from Ollerton to Edwinstowe. It was
placed about four and a half feet from the ground, and
was constructed of dry bents, intermingled with a little
moss, and lined with hair. When I found it, it contained
four eggs ; had it remained undisturbed, I have no doubt
they would have increased to the usual number of six, as
the female was on the nest. As it was, I appropriated
them as a valuable addition to my collection. This,
however, was not a solitary instance, for two years later, on
May 13th, 1856, another nest was taken from the same
hedge, near the place from which I had taken the previous
one ; it contained one egg, which was brought by the finder
to me. A third nest was taken the next day at Ollerton ;
it was placed in the side of a cattle hovel, amongst the
thorns with which the upright framework was interlaced,
and was constructed of dry grass only, and lined, as were
the others, with hair.
"The second nest had moss mixed with the grass, like the
first." It is singular that these nests were placed in situations so different from those authors describe as usually
frequented by them."
Mr. Harting, who kindly referred me to the foregoing
instances of the taking of the nest of the Black Redstart,
has published his belief ('The Field,' 1874, p. 200) that
the specific name of this bird is derived from the Greek
adjective τίθος (domesticated), the bird being frequently seen
perching on house-tops and garden walls, and building in
holes and crannies in the neighbourhood of man's dwelling.

REDSTART.

RUTICILLA PHŒNICURUS, *Linn.*

Pl. VIII., fig. 12.

Geogr. distr.—Generally distributed over the Continent of Europe; tolerably common in England, where it arrives in April, leaving again in September.

Food.—Insects, larvæ, spiders, small worms, grubs, and berries.

Nest.—Either loosely or firmly constructed of dried grass-stems, fine roots, and moss lined with hair and feathers.

Position of nest.—In hollow trees, holes in roofs of buildings, or in walls; other eccentric shelters, such as are frequently chosen by the Robin, are also sometimes selected, as, for instance, an inverted flower-pot.

Number of Eggs.—5-8.

Time of nidification.—IV-VII; May and June.

The nest of the Redstart is not easy to find, and I have never yet had the pleasure of taking it. This bird appears to be fond of the habitations of men, and consequently its young often fall victims, as do those of the Robin when nesting in similar situations, to the rapacity of the domestic tiger; "it breeds regularly in all the counties of England and Wales, though very rarely to the westward of Exeter. In Scotland it is found in Summer as far as Caithness, but does not occur in the Hebrides, and only occasionally in Shetland, and everywhere seems to be less numerous than in South Britain. In Ireland it is also rare."*

* Yarrell's 'History of British Birds,' 4th edition.

STONECHAT.

PRATINCOLA RUBICOLA.

Pl. VIII., figs. 13-15.

Geogr. distr.—Central and milder parts of Northern Europe to Northern Africa ; also India, China and Japan. Resident in Great Britain ; common in the West and South of England, in Scotland and Ireland.

Food.—Insects in all stages, and worms.

Nest.—Formed of hay and moss, lined with hair, fine bents, and feathers, and occasionally with wool.

Position of nest.—Concealed in a collection of whins, under a low bush or shrub, on bush-covered heaths or furzy commons.

Number of eggs.—4-6.

Time of nidification.—IV-VII ; May to June.

This species is very shy, and its nest difficult to discover, but the collector who has noticed the presence of the bird may sometimes obtain the nest by watching the hen bird from a distance, when she leaves her eggs to feed ; for this purpose a field-glass is useful.

The Stonechat is very active in its movements ; it is fond of perching on the topmost twigs of hedges and bushes, darting away as one approaches ; the first time that I saw this bird sitting on the top of a furze bush, darting out or down after a fly or other insect, and returning to the same spray, I concluded that the nest was under that very bush ; but as I approached the bird flew far away, settling upon a bush at a distance ; nevertheless I wasted some time in vain search.

The nest of the Stonechat has been found with eggs before the middle of April ; it occurs in every county of Great Britain, though rare in Orkney and Shetland.

WHINCHAT.

PRATINCOLA RUBETRA, *Linn.*

Pl. VIII., figs. 16, 17.

Geogr. distr.—Occurs in Persia, which it leaves in the autumn for North Africa and South Europe; it is found throughout Europe, its range extending further northwards than that of the Stonechat.

Food.—Insects, larvæ, small worms, and berries.

Nest.—Rather large, formed of grass bents, fibrous roots, and sometimes a little moss; lined with fine bents and hairs.

Position of nest.—On the ground, carefully concealed amongst grass or bushes on pastures, especially on undulating ground; in a furze bush on heaths, commons, and other open localities.

Number of eggs.—4-6.

Time of nidification.—V-VIII ; May to June.

In habits the Whinchat is much like the Stonechat; sitting, like it, upon the top spray of a furze bush watching for flies : it arrives in our country about April ; it is abundant in Somersetshire, Wiltshire, and Gloucestershire, and, according to Hewitson, in Westmoreland ; it, however, breeds with more or less regularity in all the counties of Great Britain ; it rears two broods in the season : it is not, however, a common bird, and, unless the collector frequently shifts his hunting ground, he may be years before he discovers a nest.

COMMON WHEATEAR.

SAXICOLA ŒNANTHE, *Linn.*

Pl. VIII., fig. 21.

Geogr. distr.—All over the Western Palæarctic Region from Greenland to Africa, and eastward through Siberia to North China; also
occurs in Eastern N. America and Behring's Straits; common, but
local, throughout the British Isles, arriving late in February and
leaving in September.

Food.—Insects in all stages, worms, and Mollusca.

Nest.—Made throughout of very fine dried grasses, mixed with
small fragments of wool or moss, with feathers and hair; rather
large and flat.

Position of nest.—In any hole in a wall or bank, in a rabbit-
burrow, or behind a large clod or stone, sometimes in a heap of stones.

Number of eggs.—5-8.

Time of nidification.—IV-VI; April or beginning of May.

Though it is certain that a considerable number of
Wheatears must breed in this country every year, I have
hitherto only met with the nest once in Kent; it was placed
in a hole in the side of a bank enclosing a watercress
stream, in which place it had bred for several successive
years; when I saw it no eggs had been deposited. The
bird is esteemed a delicacy for the table, and numbers are
destroyed for this purpose.

" When the nest is in a rabbit-burrow it is not unfrequently visible from the exterior, but when under a rock
it is often placed a long way from the entrance, and out of
sight. It can nearly always be found with certainty by
watching the hen-bird, and Salmon says that on the large
warrens of Suffolk and Norfolk its position is easily detected
by the considerable number of small pieces of the withered
stalks of the brake amassed at the entrance of the burrow.
When the place of concealment, however, is beneath a
rock or earth-fast stone, the nest is often inaccessible to the
finder." (Yarrell, 4th ed.)

A writer in the ' Field ' for April, 1871, suggests that the
name of this bird may have been derived from its note,
Wheet-jur, or possibly from the white base to the tail, *ear*
being a transposition of the old word *are*; the former
suggestion seems to me the least fanciful.

Sub-family Sylviinæ.

REDBREAST.

Erithacus rubecula, *Linn.*

Pl. VIII., figs. 7-10.

Geogr. distr.—Throughout Europe, visiting Algeria, Lower Egypt, and Palestine in winter; also found in Persia, Madeira, the Canaries, Azores, and Teneriffe; common and resident in Great Britain.

Food.—Worms, insects in all stages, seeds, berries, and fruits.

Nest.—Strongly and compactly built when necessary, but when in holes carelessly put together, formed of fine roots, bass, or coarse, dead grass, bents, hair and moss, and lined with hair fibre and fine grasses; a few oak leaves are frequently interwoven in the outer walls, which render it in a measure like the nest of the Nightingale; when built in holes a good deal of moss is used.

Position of nest.—In holes in grassy banks, in deserted chalk-pits, or at the side of a road; in holes or crannies in rocks, walls, outhouses, dust-bins; in holes in trees or in the ground at the foot of a tree or ivy-grown stump; in flower-pots, watering pots, pewter quart pots, hanging on a nail or fence; in ivy on a wall; in the side of a bean-stack; in fact, almost anywhere where a nest can find a secure lodgment without the necessity for strong attachment to its environment, as in nests built in trees or bushes.

Number of eggs. - 4-6; rarely less than 5, occasionally 7.

Time of nidication.—III-VII; May and June.

The Robin is one of the most useful and attractive birds; its food consisting chiefly of worms and insects, to obtain which it will follow one about, and even perch upon the spade with which one is digging; indeed, so fearless does it become if allowed to be familiar, that I have sometimes been afraid of injuring it when gardening; it appears to like to be talked to, and shows signs of impatience when taken no notice of. In choosing a place for its nest the Robin certainly prefers the habitations of men to the open country, though the nest is abundant enough in holes in high grassy banks skirting woods, particularly where scraps of ivy partly cover the soil, or where there are projecting moss-grown stumps of decayed trees.

NIGHTINGALE.

DAULIAS LUSCINIA, *Linn.*

Pl. IX., figs. 8, 9.

Geogr. distr.—Throughout Western and Central Europe ; common in S. Europe, migrating to Africa before the winter; in Great Britain it appears to be restricted to England, arriving early in April.

Food.—Insects in all stages, spiders, worms, woodlice.

Nest.—A large and bulky cup-shaped structure, loosely put together ; of dead leaves, usually of the oak, within which are coarse, dry, flattened bents, rushes or even fine flags, lined with finer bents, root fibre, and sometimes a little horsehair.

Position of nest.—Usually in a depression in the ground, well concealed by ferns, grasses, or other short undergrowth at the foot of a tree, pollard, or bramble bush; less commonly in the forking base of a pollard overhung by fern fronds and rank grass, or even in the open places in woods, groves, plantations, gardens, and hedge-rows the nest is rarely found above the ground level.

Number of eggs.—4-6.

Time of nidification.—V-VI ; end of May.

Although the nest of the Nightingale is almost invariably placed upon the earth, I met with it, about the year 1875 or 1876, in a stunted hawthorn in Herne Wood, near the village of Herne, Kent, at a height of about two feet from the ground. Again in 1882 (as recorded in the ' Zoologist ' for the year following), I found a nest " built fully eighteen inches from the ground in a matted bush of furze and bramble " near Sittingbourne, in Kent. The first of these was only just completed when I found it, the second contained a single egg. This nest, like many others, is generally discovered by the sudden flight of the bird as one approaches ; the sound made by the wings of a startled bird of about this size need be only once heard to be recognized ; it may be represented by the word *fferrelup*, and this word in bird language signifies, almost invariably, " my nest is less precious than my own safety."

On the wing the Nightingale has a bright russet-red appearance ; the flight is quick and flurried, but where it is common the bird may be often seen, both singing upon the branch of a tree and on the wing flying in and out of the skirts of a wood.

The song is the most beautiful combination of sounds uttered by any of our songsters, and to listen to five or six of these birds in early morning, or at twilight, in the dense Kentish woods, has always been one of my chief pleasures in the spring-time. Impossible as it is to express in words the music of the Nightingale, which combines the sweetest and liveliest with the most plaintive notes of our favourite singing birds, there is always one part of its song by which it can be recognised, commencing with a long-drawn plaintive *fwee, fwee, fwee, fwee,* repeated four or six times in succession, and succeeded by a chuckling rapid *chookachookachookachookachookachookee,* which has been called by poets the ' jug-jug of the Nightingale ;' how such an unpoetical expression of the sound came to be applied to it I do not pretend to understand, the chuckling does not even terminate so abruptly as in my attempt to express it, but passes off into other more pleasing sounds.

The nest, even when placed in the most exposed situations, may easily be passed by anyone not specially seeking for it, as it merely looks like a round hole in the earth surrounded by many dead leaves, and containing a few pebbles ; it varies greatly in depth, but is always rather deep, so that at times it requires a close inspection to see the dark-coloured eggs at the bottom ; these are generally of a deep olive or stone-brown colour of the type represented by my fig. 9 ; the zoned variety, fig. 8, is much rarer, and I have only once taken a clutch of this type. In its lowly position, though not a conspicuous object to us, the nest is subject to the attacks of stoats, rats, and other wood-infesting vermin, and therefore it is not very unusual to find a clutch of broken egg shells in place of eggs.

BLACKCAP.

SYLVIA ATRICAPILLA, L*inn.*

Pl. IX., figs. 10, 14.

Geogr. distr.—Throughout Europe, excepting in the extreme north; eastward into Persia, and in Africa as far southward as the Gambia; also found in the Canaries, Madeira, and the Azores; in Great Britain it arrives in April and leaves in September or October, and is to be met with commonly in England and Wales.

Food.—Insects in all stages, spiders, Crustacea, berries, and small fruits.

Nest.—Cup-shaped, formed of dry grass, straws and root fibre, with occasionally a little moss; lined with fine bents and a few horse-hairs.

Position of nest.—In brambles, creepers, bushes or dwarfed trees in small woods, groves or gardens where there is plenty of dense undergrowth; also rarely in the outskirts of dense forests.

Number of eggs.—4-5.

Time of nidification.—IV-VII; beginning of June.

I have usually found the nest of this bird in a straggling and confused mass of bramble, honeysuckle and stinging-nettle, either in small open places in woods and groves, or else close to a narrow path just within the entrance to a wood. I believe it to be useless to look for this nest, excepting where the undergrowth is tolerably dense, as it almost invariably is in the Kentish woods, where one frequently has to find one's way out backwards in order to make an exit after a morning's collecting. As the male usually does the incubating, it is not so difficult to recognise the nest when found as one would imagine.

The eggs vary in tint considerably more than those of the Garden Warblers; the red variety (Pl. IX., fig. 14) is one of a clutch which I took at Tunstall, in Kent, on May 24th, 1877, and on the 29th of the same month in 1878 I took a second clutch of five eggs, slightly larger but similarly coloured, within a hundred yards of the place where the first was obtained; if these were laid by the same bird, as seems likely, it would tend to show that the red colouring was confined to individuals of the species.

GARDEN WARBLER.
Sylvia salicaria, *Linn.*
Pl. IX., figs. 15, 17.

Geogr. distr.—Europe up to 69° N. lat. in summer; S. Africa in winter; in Asia extends as far east as the Ural range; common in England, and in Scotland as far as Banffshire; rarer in Ireland and Wales; arrives in April or May, and leaves in September.

Food.—Insects in all stages, spiders, berries, fruit, &c.

Nest.—Much like that of the Blackcap, but more slightly put together; it is formed of goose-grass and other fibrous plants, with sometimes a little moss and wool, lined with fine roots and a few hairs.

Position of nest.—In low bushes and brambles in gardens and copses.

Number of eggs.—4-5.

Time of nidification.—V-VIII; June.

The song of this bird is wild, rapid, mellow, but irregular; the nest is almost as abundant as that of the Blackcap, and towards the end of May or beginning of June it may be confidently sought for amongst the undergrowth in thickets and narrow woods; in such places I have taken a considerable number of their eggs, none of which, however, bear out Hewitson's opinion, that eggs of this species occasionally have the rich colouring of the Blackcap; they have, indeed, been always coloured more or less like the three varieties figured on my plate, of which fig. 15 is the most usual form and fig. 17 a rare albino variety; when eggs of the Blackcap approach them in colouring, they are generally to be distinguished by their slightly superior size and a sprinkling of dark spots or splashes: this, however, is only the experience gained in about fifteen years' collecting, almost wholly in one county, and therefore cannot be considered conclusive; at the same time I have never passed by a single variety which I was not certain I possessed, and have, therefore, brought together a selected series far superior to many that I have seen in larger collections.

The nest may be found, as Hewitson says, at the same season of the year as that of the Blackcap; but I have always found it about a fortnight or three weeks later.

WHITETHROAT.

SYLVIA RUFA, *Bodd.*

Pl. IX., figs. 18, 21.

Geogr. distr.—Throughout Europe up to about 65° N. lat.; also found in Western Asia; arrives in Great Britain in April or May, but migrates to N. Africa in September.

Food.—Insects in all stages, berries, peas, currants, &c.

Nest.—Lightly constructed of dried stalks of plants and grasses, with here and there little woolly knots of spiders' web, lined with fine bents and horsehair.

Position of nest.—Low down in small loose bushes, brambles or nettles.

Number of eggs.—4-5 ; rarely 6.

Time of nidification.—V-VI. May.

The nest of this species is, as a rule, wonderfully uniform in construction ; in a series of ten nests taken by myself during two consecutive years only two are worthy of note ; one of these is chiefly remarkable for its unusual size, the diameter of the interior of the cup measuring nearly three inches (about three-quarters of an inch more than usual), and being thickly lined with black hair ; the other nest is rather thickly edged with pieces of sheep's wool twisted into the grasses ; at the same time there is considerable difference in the strength of the nests, for, although most of them are loosely put together and extremely light, I have met with them as thick and firm in the walls as the majority of the nests of the Sedge Warbler, and, where both birds are breeding in the same cover, it is necessary to be wide awake so as not to mistake some varieties of the eggs of the Sedge Warbler for those of the Whitethroat ; there is, however, rarely any difficulty in distinguishing them.

Unless the eggs are partly incubated, the parent bird usually slips off them very quietly as one approaches, and flies for some distance near the ground before rising, so that I have, when not expecting to meet with the nest, on more than one occasion imagined that I had seen nothing more than the hurried departure of a field mouse ; when flushed, however, its flight is sudden and startling.

LESSER WHITETHROAT.
SYLVIA CURRUCA, *Linn.*
Pl. IX., figs. 22, 24.

Geogr. distr.—Generally in Europe during the summer, arriving late in March, in April, or, according to Newman, as late as the beginning of May, but retiring before the winter to Africa; occurs in Asia as far east as Dauria and China; not uncommon in Great Britain.

Food.—Insects, berries, and small fruits.

Nest.—A neat and very compact, though not always strongly-built, cup-shaped structure, of stout bents intertwined with a few rootlets, a little fine wool, and spiders' web, and lined with fine bents, rootlets, and a little horsehair.

Position of nest.—In uncut bushes, tangled brambles and hedges on the outskirts of woods, in groves or shrubberies.

Number of eggs.—4-5.

Time of nidification.—IV-VI; May.

I have always found this a shy bird, requiring a very slight inducement to desert its nest; on two or three occasions when I have found the nest to contain only a single egg, and have substituted a small marble sufficiently like to deceive most birds, I have subsequently found the marble ejected and the nest deserted by the bird, no second egg having been laid. The nest is a much firmer, and usually smaller, structure than that of the common White-throat, and the eggs are so distinct that there is no danger of mistaking them for those of any other British bird; for, though a little like some eggs of the Blackcap in colouring, they are much too small to be taken for them.

The nest appears to be comparatively rare, for, though I generally come across two or three in a season, they are not unfrequently deserted after the first egg has been laid; it even appears as though this species were almost as jealous of the discovery of its nest as the Wren; once or twice when I have again visited a nest after the lapse of a week I have found only the second egg deposited; it more frequently happens, however, that, owing to its conspicuous position, the nest has been pulled out and trampled on by some village urchin.

DARTFORD WARBLER.

MELIZOPHILUS UNDULATUS, *Bodd.*

Pl. X., fig. 4.

Geogr. distr.—Western Europe as far to the northward as Great Britain; otherwise confined to the south; N. Africa, especially the western portion. In Great Britain it is tolerably common, though local, in the southern counties; excepting in Middlesex, it does not breed further north than the Thames; not known in Scotland or Ireland; it is not rare in heathy parts of Kent, Surrey, Sussex, Hampshire, Wiltshire, Dorsetshire, Devonshire, and Cornwall.

Food.—Small insects in all stages.

Nest.—Loosely constructed of stems of goose-grass and other vegetable stalks and young dry furze mixed with wool and lined with a few dry stalks of *Carex*; altogether not unlike the nest of the Greater Whitethroat.

Position of nest.—In dry furze bushes.

Number of eggs.—4-5.

Time of nidification.—VI-VII.

I have not hitherto had the pleasure of discovering the nest of this species, and the egg which I have figured is from Mr. Seebohm's collection; as will be seen, it is decidedly darker than that of the Whitethroat, and quite unlike the egg figured by Hewitson, which would do equally well for that species. The nest appears chiefly to differ in having furze mixed with the grass-stalks which compose its walls, a combination which I have never found in the nest of *Sylvia rufa*. According to Newton (Yarrell's Brit. Birds) nests built early in the season are more compact, and rather resemble those of the Sedge Warbler, and such certainly is the impression conveyed by the admirable representation of the nest at the end of his article.

Sub-family *ACROCEPHALINÆ.*

GRASSHOPPER WARBLER.

LOCUSTELLA NÆVIA, *Bodd.*

Pl. VIII., figs. 19, 20.

Geogr. distr.—Central Europe; occurring as far north as Scotland in Summer; here it is more local than in England, where it breeds in every county, as also in all suitable localities in Ireland and Wales.

Food.—Snails, slugs, worms, larvæ, and insects.

Nest.—Cup-shaped; strongly and neatly formed of grass-bents and moss, with occasionally a few leaves; lined with finer bents.

Position of nest.—On the ground, carefully concealed in herbage in bush-covered localities and hedges.

Number of eggs.—5-6; rarely 7.

Time of nidification.—V-VI.

The nest of the Grasshopper Warbler is difficult to find, as it is usually placed deep down in the centre of a tuft of coarse grass, so that there is nothing externally to show that the tuft has been thus taken possession of; when disturbed the hen bird slips quietly out of the tuft, and works its way through the surrounding herbage. In the early days of my collecting I remember to have observed this trick, and to have wondered that I could not find the nest; the reason was simply that I did not go deep enough for it: this nest has, however, been discovered at the bottom of a ditch overgrown by coarse grass and the prickly branches of a whin-bush, through which the bird was seen to enter: all of this had to be removed before the prize could be taken possession of.

SEDGE WARBLER.
ACROCEPHALUS SCHÆNOBÆNUS, *Linn.*

Pl. IX., figs. 1-4.

Geogr. distr.—Throughout Europe, migrating to N. Africa towards winter; it visits the British Isles in April and leaves them in September.

Food.—Insects in all stages, worms, slugs, and snails.

Nest.—A somewhat deep and usually compact cup; constructed most frequently of dried grass, fine rootlets, and moss, lined with horsehair, feathers, and sometimes a little wool.

Position of nest.—Low down in the outer branches of thorn bushes or hedges and brambles, sometimes overhanging dykes and ponds, in sedges and reeds, or among the stems of water-plants on swampy ground.

Number of eggs.—4-8; usually 5.

Time of nidification.—V-VI.

This species breeds everywhere in Great Britain; I have usually found its nest in thick hawthorn bushes overhanging water: certainly the most pleasant position in which to discover it, as it enables one to return home dry-shod; Mr. Dresser, on the contrary, seems to have usually obtained it on marshy ground in patches of aquatic herbage, especially where *Euphorbia*, &c., are plentiful rather than where reed predominates; in Norfolk I have taken it out of the reeds and sedges which border some of the dykes or surround the islands on the Broads; it, however, also occurs amongst nettles in wild places, and is then usually made with withered goose-grass loosely put together, and is shallower than when placed amongst reeds. I have, moreover, taken the nest in woods, amongst brambles, in just the same position as that occupied by the Garden Warbler, and my friend Mr. Salter took one or two nests out of ordinary roadside hawthorn hedges near Salisbury.

The commonest type of egg is that represented by fig. 1, but some eggs are indistinguishable from those of the commoner yellow Wagtail, excepting that they usually have a little black line like a crack at the larger end; fig. 2 represents a very aberrant variety, approaching some eggs of the Reed Warbler.

REED WARBLER.
ACROCEPHALUS STREPERUS, *Vieill.*

Pl. IX., figs. 5-7.

Geogr. distr.—In Europe as far northwards as S. Scandinavia; it winters in Africa, and in Asia ranges as far east as Turkestan; has a more restricted range in the British Isles than the Sedge Warbler, and is decidedly rare in Scotland and Ireland.

Food.—Insects, worms, slugs and snails.

Nest.—A deep cup formed of dried grasses and bents, or the flowering tops of the reed; sometimes a little moss and a fair sprinkling of cobweb; lined with fine grassy fibre.

Position of nest.—Interwoven, sometimes loosely, sometimes firmly, with two or three growing reeds, usually with three, in millpools, broad dykes, fens, and reedy banks of rivers and broads; also in hazel trees near water.

Number of eggs.—5-6.

Time of nidification.—V-VI; end of May.

I have found that when anywhere in the neighbourhood of houses the best method of obtaining this nest is to let a long ladder fall from the bank across the reeds, which will then support it upon the surface of the water so securely that it is possible to walk out upon the rungs, almost to the extremity, nearly dry shod; this enables one to look right and left through the growing reeds, and to reach the nests, which are rarely placed near to the bank. Where there is considerable ebb and flow the nest is not invariably placed above high-water mark, but is made additionally thick at the bottom, and woven loosely around the reed-stems above a leaf, so as to rise upon the surface of the water when at its highest; such was indeed the case with a nest (which, by the bye, contained a cuckoo's **egg**) taken by myself in Kent on the 5th June, 1875; the eggs in this nest were hard-set, but others obtained at the same time were quite fresh. I have three nests of this species built in forks of hazel; the first of these is normal in construction, and was obtained for me by my friend the Hon. Walter de Rothschild at Tring; the second was sent to me from Salisbury by Mr. Salter, is unusually large and compact, formed of carefully-selected stout grasses interwoven with some apparently vegetable woolly substance, and bound tightly round externally with stronger grasses; it contains four eggs, decidedly larger than usual, and resembling one of the varieties of the eggs of the Marsh Warbler, though not one so strongly marked as my figure. When I received this nest I was fairly puzzled, for it answered tolerably well to the description of that of the Great Reed Warbler, the eggs to some of those of the Marsh Warbler, the position of the nest to that which I then knew to be occasionally adopted by the Reed Warbler; I therefore wrote to Mr. Salter asking him to try and discover something further respecting it. On the 27th June he wrote to me from Downton as follows:—" Dear Sir, I will forward, per parcels post, to you, another nest like the one you have. I found it last Saturday with three young birds and one egg. I went again to-day and found the young ones just ready to fly. I managed to shoot one of the old ones with a catapult, but could not manage to get the other although I waited about three hours. The nest was overhanging the water about fifty yards from where I got the other."

The egg contained in the second nest (which was formed like that previously received) was perfectly normal, and the birds, upon comparison with a series of skins, proved to be quite typical specimens of the ordinary Reed Warbler, showing how careful we ought to be in attempting to identify nests or eggs without seeing the birds, although, of course, puzzles of this kind do not trouble the collector very frequently.

"The Reed Warbler," says Mr. Harting (' Zoologist,' 1867) " may be distinguished from the Sedge Warbler by its being longer and slimmer, and by the uniform colour of the head. In the Sedge Warbler the most conspicuous characters are a white line over the eye, a darker back, and dark centre to wing feathers, with lighter margins. In the Reed Warbler the feathers are more uniform in colour. The two species differ also in their note and flight."

MARSH WARBLER.
ACROCEHPALUS PALUSTRIS, *Bechst.*
Pl. XL, fig. 2.

Geogr. distr.—Continental Europe in the summer; Asia as far east as Persia; winters in N. and perhaps S. Africa; in England it has occurred several times, and lately several instances of its breeding in this country have been recorded.

Food.—Insects and soft fruits.

Nest.—Formed of dry stalks and leaves of grasses, mixed with nettle fibre and webs, and lined with fine bents and few horsehairs, or moss and a quantity of horsehair.

Position of nest.—On the edges of dense thickets near to ditches and moats, suspended from one to three feet from the ground in isolated little bushes, nettles, fig-wort, the greater willow herb, cow-parsnip, or in meadow-sweet.

Number of eggs.—5-7.

Time of nidification.—VII.

The breeding of this species near Taunton, in Somerset-shire, was recorded by Mr. Cecil Smith in 1875, and Mr. Seebohm (Hist. Brit. Birds, vol. 1, p. 375) states that in 1882 three nests were taken in the same locality, two of which are now in his collection; he also records the finding of two nests near Bath by Mr. John Young; the following description of the nest I quote from p. 378 of Mr. Seebohm's book:—

" The nest perfectly resembles that of the Grasshopper Warbler, but is closer built, and its colour is darker and greyer; it is also more smoothly finished outside. It is as deep as the nests of other Reed Warblers, neatly rounded, with the upper edge bent inwards. The materials are principally dry leaves and stalks of fine grass, mixed with grass and the fibres of nettles and other plants, and often with insect webs, all somewhat carefully woven together, in some places almost felted together. Inside it is lined with very fine grass and a considerable quantity of horsehair.

" The two nests from Taunton were suspended between stems of the meadow-sweet."

The egg figured on my plate was lent to me by Mr. Seebohm.

Though very like the Reed Warbler, Mr. Seebohm says that freshly-moulted birds of the Marsh Warbler can be distinguished from that species by the russet-brown instead of olive-brown colour of the rump.

SUB-FAMILY *PHYLLOSCOPINÆ*.

WOOD WREN (OR WARBLER).

PHYLLOSCOPUS SIBILATRIX, *Bechst.*

Pl. X., figs. 1, 2.

Geogr. distr.—In Europe extends northward as far as S. Scandinavia, and eastward to the Ural Mountains ; winters in N. Africa, breeding generally in England, Wales, the Southern and Midland Counties of Scotland ; in Ireland it is rare.

Food.—Insects and larvæ.

Nest.—A domed structure with stout walls, composed of dry grass bents, a few dead leaves and a little moss, neatly lined with horsehair and finer bents.

Position of nest.—On the ground, concealed amongst dry foliage or a tussock of grass, usually in beech or oak woods.

Number of eggs.—5-7 ; usually 6.

Time of nidification.—V-VI ; early in May.

This bird is with us from about the end of April to September ; it appears to affect localities which abound in high trees, but is everywhere very local. Its song is loud, and has been represented by the word " twee " sounded very long, repeated at first slowly but afterwards more rapidly, and interrupted at intervals by the variation " chea " " chea " " chea." As with other Warblers, the song ceases as soon as the young ones are hatched.

According to Mr. Harting, the Wood Warbler differs from the Willow Warbler and Chiffchaff in its brighter green colour above and purer white beneath, the more distinct yellow line over its eye, shorter tail, and proportionately longer wings, whilst the song of all three differs enough to enable one to distinguish them at a good distance.

WILLOW WREN (OR WARBLER).

PHYLLOSCOPUS TROCHILUS, *Linn.*

Pl. X., figs. 5, 6.

Geogr. distr.—Europe generally; in Asia eastward to the Yenesay, and in Persia; in N.W. Africa, and in winter southward to the Cape; in Great Britain it is generally distributed and common in the summer, arriving towards the end of March or early in April, and leaving again in September.

Food.—Insects and berries.

Nest.—Semi-domed and externally somewhat carelessly constructed, though neatly formed inside; of dry grass, fern, dead leaves or moss, compacted with cobweb and lined with wool, hair, and feathers.

Position of nest.—Usually on the ground amongst long grass, or in a depression in the earth; or against a grassy bank.

Number of eggs.—4-7.

Time of nidification.—IV-VI.

This species occurs in groves of mixed trees, orchards, gardens, and the borders of woods, nesting chiefly in clearings or on the margins of woodland paths. A curious unfinished nest (*i. e.*, without the dome-like covering) was obtained by me in 1883; the absence of the dome was partly compensated for by an overhanging clod. This nest, which contained four unusually well-marked eggs, was placed on the ground, under a gooseberry bush in a Kentish orchard; I have described it more fully in the ' Zoologist' for October, 1883, p. 491.

Though commonly placed upon the ground, the nest is not invariably found in that position : one that I took on the 16th June, 1881, was built over two feet from the ground in the drooping branches of a wild rose-bush in a garden at Tunstall, in Kent; and in the 'Zoologist' for 1878, p. 351, E. P. P. Butterfield records two instances observed by him,—one in 1876, in which the nest was built between two rocks at a distance of three feet from the ground, and one in 1878, in which it was placed in a clump of whins two feet from the ground.*

Though the number of eggs laid by this species is said to vary from 5 to 7, it is a singular fact that I have hitherto never found more than four in a nest. Mr. Newton, in the fourth edition of Yarrell's ' British Birds,' says, " The eggs are six or seven in number."

---

* An instance is recorded by Mr. Alston of this bird building in a hole in a wall nearly seven feet from the ground.

E

CHIFFCHAFF.

PHYLLOSCOPUS COLLYBITA, *Vieill.*

Pl. X., figs. 7, 8.

Geogr. distr.—N. Europe during the summer, extends eastwards to Persia; winters in S. Europe and N. Africa; occurs also in the Canaries and Teneriffe; arrives in Great Britain about the end of March or beginning of April, often remaining with us as late as October; occasionally resident.

Food.—Insects in all stages.

Nest.—See frontispiece; an exact copy of a nest which I took in a wood near Newington, Kent, on the 12th May, 1882.

Position of nest.—On or near the ground in a bank, or behind a stump amongst tolerably dense grass or weeds; rarely above the ground; in open spots in woods, or at the edge of a grove where the ground is scantily covered with tangled herbage.

Number of eggs.—5-6.

Time of nidification.—V.

I have never met with a nest containing eggs later than May; that represented on the frontispiece was found close to a narrow path about 500 yards from the entrance to a wood, and in a small open spot densely covered with tangled, dead, reedy grass; it rested upon the top of a short mossy stump, with its back to the path, and was so closely interwoven with the dead grasses that, unless I had found the bird building it a week previously I should have overlooked it entirely. A second nest in my collection was found by Mr. O. Janson in a cavity in a steep bank at the edge of a narrow, tangled thicket, such as in Kent is popularly known as a " shore," or " shave."

The song of the bird is singularly monotonous, consisting of an incessant repetition of the same note twice uttered with the regularity of a pendulum, " chiff-chiff," not " chiff-chaff," as one would suppose from its name.

In one of the earlier volumes of the ' Zoologist' it is stated that a low bush, frequently of furze, appears to be a favourite locality for the nest; I have not personally met with it in such a position, but I should not be surprised if I did.

GOLDEN-CRESTED REGULUS (OR WREN).
REGULUS, CRISTATUS, *Koch.*
Pl. X., figs. 14, 15.

Geogr. distr.—Throughout Europe, N.W. Africa, and Asia as far east as Japan. Found in all suitable localities in Great Britain; a resident species.
Food.—Insects.
Nest.—Cup-shaped, very thick and solid, formed of green moss and lichens compacted together with spiders' web; warmly lined with feathers, especially the red breast-feathers of the Linnet.
Position of nest.—Almost invariably among the pendant twigs immediately under the branch of a fir or yew tree ; but, according to Hewitson, sometimes upon a branch, or against the trunk.
Number of eggs.—7-11.
Time of nidification.—III-IV.

The Goldcrest has been known to rear two broods in the same nest ; its habits are not unlike those of a Tit ; its food consists merely of the small insects which hide themselves between the leaves or under the bark of firs and and larches ; its song is soft, low, and pleasing ; the nest is not unlike some nests of the Chaffinch ; it is, however, a little deeper, the interior of the cup is decidedly less compact and neat, and its unusually pendent position gives it a very different aspect.

Although partially resident in Great Britain, vast numbers visit our coasts in autumn, and such of them as survive leave us again in the spring.

When sitting the hen bird is so fearless that she permits close observations to be taken without leaving the nest. The account of a nest containing young, which Montagu took into his house and watched day by day, the hen feeding the young even whilst he held the nest in his hand, has been often repeated, and proves that the maternal instinct is fully as strong in the Goldcrest as in the Blue Tit.

FAMILY ACCENTORIDÆ.

HEDGE SPARROW (OR ACCENTOR).

ACCENTOR MODULARIS, *Linn.*

Pl. VIII., figs. 4-6.

Geogr. distr.—Europe, ranging eastward as far as Persia; generally distributed, resident and abundant throughout Great Britain.

Food.—Seeds of weeds, and insects in all stages.

Nest.—Cup-shaped, rather deep, usually in a framework of hawthorn or other twigs, sometimes dead fragments of furze, the walls thick and loosely formed of a quantity of green moss, often, but not always, with a liberal intermixture of dry grass, and occasionally a little sheep's wool; thickly lined with different kinds of hair and a quantity of fine wool.

Position of nest.—In hedges, thickets, furze and other bushes; very rarely in a tuft of grass on the ground.*

Number of eggs.—4-6; usually 5.

Time of nidificaton.—III-VI; May.

This species has been known to build and lay its eggs as early as the beginning of January; but, of course, this was owing to the mildness of the weather, which not unfrequently deceives our resident birds; as a rule, it commences nidification towards the end of April, but, owing to the violent storms of wind and rain which not unfrequently occur during that month, its nest becomes often so uncomfortable that I have at that season found it deserted with its full complement of eggs and even with fledglings. May is, therefore, the month during which nests of the Hedge Sparrow most abound, and so different are some of these in external appearance that they would, at first sight, hardly be supposed to be built by the same species; thus of seven nests which I have retained for my collection one is as round, neat, and soft as the typical artist's ideal, the outer walls being formed of very fine bents, fibre and moss, whereas another is as rough as if dragged together with a rake, the outer walls being composed almost wholly of the roots of couch grass (twitch). As a rule the nest is about as neat as that of the Greenfinch. The eggs vary very little, those which I have figured representing the extreme modifications which I have met with; the pear-shaped form is, however, rare.

---

* An instance of a nest being built in a cabbage was recorded by A. E. Shaw in 1877.

FAMILY CERTHIIDÆ.

COMMON CREEPER.

CERTHIA FAMILIARIS, Linn.

Pl. X., fig. 3.

Geogr. distr.—Generally in Europe ; also Asia, N. Africa, and N. America ; in Great Britain generally distributed and resident.

Food.—Insects.

Nest.—Formed of small twigs, fine grass straws or strips of bark, and a little moss, neatly lined with feathers, and sometimes a little wool.

Position of nest.—Under eaves of straw thatches ; behind the loose plaster of a wall or pieces of partly detached bark, or trellis work ; in piles of timber, holes in trees, and other suitable situations.

Number of eggs. 5-6 ; probably generally 6.

Time of nidification.—IV-V.

Owing to the position of the nest of this bird it is usually compressed and rather deep. The Creeper is one of our smallest but most interesting birds, and it is impossible to catch sight of it, as it moves up the trunk of a tree spirally in little sharp jerks, without wishing to see more of it ; this, however, is not an easy matter, for as one follows its movements it suddenly slides out of sight, so that it is only by dodging that one is sure of seeing it again ; as it moves in this manner over the surface of the bark it is constantly watching for insects hidden in the crevices ; these it picks out with its long, slender bill. When the examination of one tree is completed it flies to another, generally working upwards from the foot.

Though the Creeper is said sometimes to lay as many as nine eggs, it is probable that six is the more usual number ; at the same time, though I have on several occasions noticed the bird, I have hitherto not met with a nest containing eggs, and, therefore, I can only judge by what I have proved to be the case with Wrens and Tits. Newman states the number to be 6 to 8, and Dresser, I think, says 5-6. The number 9 seems to have originated with Hewitson, who, I think, was rather inclined to record extraordinary clutches as though they represented nothing unusual.

FAMILY **TROGLODYTIDÆ.**

COMMON WREN.

TROGLODYTES PARVULUS, *Koch.*

Pl. IX., figs. 9-13.

Geogr. distr.—Throughout Europe from the north of Scandinavia to Algeria; eastward to Central Asia; common nearly everywhere in Great Britain, and resident.

Food.—Insects in all stages and woodlice.

Nest.—A strongly-built and usually very neat structure, varying considerably in form and materials according to the position in which it is placed, but always cave-like (the entrance being in front or at the side, never at the top); one in my collection from a laurustinus hedge is oblong-ovate, and produced in front of the entrance into a sort of half cup, which must have rendered the interior very damp in wet weather; it is the nearest approach to a top entrance that I have seen. As a rule the variations may be summed up as follows: if in hedges or among brambles in woods, it is formed of dry plant-stalks, a quantity of soft decayed leaf mixed with a little moss, and, towards the inside, a few (perhaps three or four) feathers; the outside is completely and somewhat loosely covered with decayed or withered oak and other leaves and a little vegetable fibre. If in an open bush, sometimes when in a hawthorn hedge, if against an old tree or on the top of a stump, it is almost entirely constructed externally of green moss; sometimes, again, when made against a young and vigorous tree, it is formed almost entirely of the stalks and leaves of dead grasses; if in a barn, of straw; if in hanging brambles, filled with rubbish, of dead grasses and moss. All these forms of nest I have taken recently, and others, formed almost wholly of clover or other materials which were handy, have been described; all nests, however, so far as I know, have a little, though sometimes a very little, moss and a few feathers in the inner lining; most of them are round or oval, with the entrance in front and near the top.

Position of nest.—In hedges, hawthorn bushes, furze, laurels, in ivy on walls or open caves by the roadside, against trunks of trees either openly near the ground or higher up in ivy; in brambles and straggling undergrowth in woods, under overlapping ledges of steep banks, in faggot-stacks, against clover and other stacks, under projecting thatches of outhouses, upon a beam in the wall of a barn, but not in holes like the nest of the Blue Tit, which countrymen and tyros commonly mistake for that of the Wren.

Number of eggs.—Usually stated to be 7-8; the regular number is certainly 6, so far as my experience goes; and I have never found a nest with more: out of seven nests in my possession, which I took during two seasons, four have a full clutch of 6, and in the others the clutch is incomplete; the same may be said of the whole of those found by me during eleven consecutive years.

Time of nidification.—V.

Not only does this species do its utmost to conceal its

nest by forming it, as a rule, of materials which render it inconspicuous on account of its resemblance to its surroundings, but it is extremely jealous of even the approach of man during the process of construction. On one occasion, whilst I was watching the building of a nest from what I thought a safe distance, the bird seemed to become aware of my presence, and abruptly deserted; even when the nest is completed, and after eggs have been deposited, the mere insertion of one's finger into the entrance hole during the absence of the parent bird will almost invariably result in desertion. Only on one occasion, in the case of a nest built under a tall drooping furze bush, I succeeded in abstracting two eggs (substituting for them small white pebbles), and subsequently, on visiting the nest, I discovered that the clutch had been completed; such instances probably occur but once in a life-time. After the eggs are hatched the mother bird is less timid, and Hewitson speaks of one that even sat still upon eggs whilst the nest was handled, a statement I could believe if the bird had been the Blue Tit, but a fact hardly conceivable in the case of the Wren. In the 'Zoologist,' Oct. 1883, I have noted the adoption of a Swallow's nest by this bird.

Family SITTIDÆ.
COMMON NUTHATCH.
SITTA CÆSIA, *Wolf.*
Pl. X., figs. 16-18.

Geogr. distr.—Central and Southern Europe ; Siberia, Persia, and Algeria ; England generally, though rare in the north-west and in Scotland ; not recorded from Ireland, but common in the north of Wales.
Food.—Insects, mast, nuts, and fruit.
Nest.—A few dry oak leaves or fragments of fir-bark, and sometimes a little grass.
Position of nest.—In holes in trees, or occasionally stacks, the entrance being plastered up with mud or clay, until just large enough to admit the bird.
Number of eggs.—5-7 ; usually 5.
Time of nidification.—V-VII ; May.

This species is resident with us, inhabiting woods, groves, and parks in the summer, but visiting orchards and gardens in the winter. It cracks nuts by fixing them in a crevice in a tree or post and striking them repeatedly with its bill ; the sound of these strokes can be heard for some distance, and from this the bird has obtained its popular name.

When sitting, the Nuthatch makes violent demonstrations in defence of her eggs or young, hissing and pecking at one's fingers with unpleasant force ; in this habit, as also in its manner of roosting with the head downwards, and in the position of its nest, it shows its affinity to the Tits.

An interesting account illustrating the affection of the Nuthatch for its young was published by Mr. J. E. Harting in the 'Field' for October, 1873, p. 348. A friend of his, Mr. William Borrer, of Cowfield, near Horsham, put up some boxes in trees near his house in March, 1871, and next week two of them were taken possession of, each by a pair of Nuthatches. On visiting one of these boxes in June or July he found only two young birds nearly ready to fly ; he then took the box indoors and left it for about an hour in the hall, the door being shut ; having got out his dogcart, he placed the box between his feet and started for Henfield, about 4½ miles distant, and when half-way there saw a Nuthatch fly over the box close to his knees; it chirped to the young birds, but no answer was heard, nor had he heard them utter any note previously. The box and birds were handed over to his sister at Henfield, and were placed in a cage under a verandah outside her bedroom window. The following morning an old Nuthatch was seen feeding the young, and the day after two old birds were there : these birds continued to visit the cage, sometimes in the verandah, and sometimes in the bedroom, for a week or two. The young birds became perfectly tame, but some months later the birds came to an untimely end. Mr. Harting adds that about three months after the young birds had been caged, namely in September, he was staying with Mr. Borrer, who related to him the above anecdote, and offered to drive him over to see the birds ; the offer was accepted, and on their arrival at Henfield they were informed that the old birds had been seen feeding the young a few days previously. After examining the young in their cage, which was a large one, they seated themselves within view of it, behind a large shrub, and about twenty minutes later, on looking out, saw an old Nuthatch clinging outside the wires of the cage, and thrusting its bill through and towards a young one which was clinging to the wires inside and opposite to it. A few minutes later a second bird appeared, and alighted on the top of the cage and peered down into it in different directions. After watching them for some time Mr. Harting and his friend stepped forward, when both the old birds flew away. They again retreated and waited, and the birds returned and behaved as before. Mr. Harting concludes as follows :—" If these birds were really the parents of the young, as is rendered probable by the incident of the old Nuthatch flying across the dogcart on the road, and not, as is possible, the adopted parents, a more extraordinary instance of intelligence and natural affection in birds I do not remember to have met with."

Family PARIDÆ.

GREAT TIT.

Parus major, *Linn.*

Pl. X., figs. 19, 20.

Geogr. distr.—Throughout the Palæarctic Region; common in Great Britain, perhaps least so in the north of Scotland; resident in England.

Food.—Seeds, nuts, insects, larvæ, eggs, and brains of small birds.

Nest.—A bed of hair, wool, or feathers, upon a thick foundation of dried grass or moss.

Position of nest.—In holes in aged forest trees, or in walls of gardens or shrubberies; sometimes behind detached planking of houses or arbours.

Number of eggs.—6-8; usually 6.

Time of nidification.—V-VI; May.

The Great Tit has been known to nest under a garden pot, the inside of a pump, or even the deserted nest of some larger bird; I have, however, found it myself only in a hole in the side of an aged tree, which was wholly filled up with the mossy mass; the eggs, six in number, are almost pure white, the spots upon them being unusually small.

The note of the Great Tit has been likened to the sharpening of a saw; it is, however, so exactly the sound made by the ungreased wheel of a loaded barrow when rapidly propelled, that as a boy I always recognised it by the name of "Wheelbarrow bird;' indeed, when I first took notice of the sound, I for some time believed that it was thus mechanically produced, and was not a little surprised to discover that it was the note of a bird. Considering how abundant the Great Tit is, even occurring in London gardens and churchyards, it is surprising that one does not more frequently find its nest.

BLUE TIT.

PARUS CÆRULEUS, Linn.

Pl. X., figs. 21-23.

Geogr. distr.—Generally distributed in Europe; common throughout Great Britain, and resident

Food.—Insects, berries, and fruits.

Nest.—A thickly-compacted layer of moss, fine fibre, dead leaves, feathers, cobweb, and a few grasses, warmly lined with feathers.

Position of nest.—In holes in walls, hollow trunks of small trees, door posts, lattice work of summer-houses, old pumps, on tops of walls under overhanging thatches and similar suitable situations; most commonly found in parks, gardens, orchards, and outhouses.

Number of eggs.—8-10; usually 8.

Time of nidification.—V-VI.

The nest of this species is commonly regarded by rustics and young collectors as that of the Wren, chiefly from the similarity of the eggs in the two species and the small size of the birds ; but, whilst the present species has almost to be lifted from its nest before it can be induced to leave it, and, after hatching its eggs, may be taken from it hissing with indignation at being disturbed, the least notice will frequently cause the Wren to desert its home altogether ; an instance may be cited in support of this statement :—

On the 27th June, 1881, I found the nest of a Blue Tit in a hole left by the removal of a brick in the wall of an outhouse, which was daily visited by a gardener ; the nest contained four eggs, one of which I removed daily, substituting a marble for the last egg ; three or four days subsequently I visited the nest and found the Tit contentedly trying to hatch out the stone ; I next removed the marble, and, later on, the nest ; but, to my astonishment, the bird even then returned for a day or two to the vacant hole in the wall. So blind and unreasoning is the instinct of this bird, that, unless I had myself tested it, I should have thought it incredible ; one may readily believe that the unsympathetic cat may become attached to a mere locality, but one looks for more sense in a bird.

I have the eggs of this bird in the nest of the Sand Martin, of which it had taken possession ; no additional lining is added.

ENGLISH COAL (or COLE) TIT.

PARUS BRITANNICUS, *Sharpe and Dresser*.

Pl. X., fig. 24.

Geogr. distr.—So far as is at present known, confined to the British Isles, where it is not uncommon, breeding in every county as far as Sutherland.

Food.—Insects, worms, seeds, berries, and fruits.

Nest.—A thick structure composed of moss and fine grass, lined with hair and wool or rabbits' fur, with occasionally a few feathers.

Position of nest.—In a hole in a trunk or branch of a tree, or occasionally in the ground in woods throughout England and Wales ; in pine forests in Scotland.

Number of eggs.—6-9.

Time of nidification.—IV-V.

Supposed to be more abundant in Great Britain than formerly, and said to be undoubtedly so in winter. I have received the nest of this species from Kent as late as the beginning of June, containing a single egg only ; it was taken out of a decayed fruit tree, the heart of which it occupied, and was at no great distance from the ground.

MARSH TIT.
PARUS PALUSTRIS, *Linn.*

Pl. X., fig. 28.

Geogr. distr.—Found in many parts of Europe, in Asia Minor and
Persia; throughout England and Wales, but scarce in some parts of
Scotland and Ireland.

Food.—Insects, larvæ, and seeds.

Nest.—Formed of moss and hair, willow or thistle down, and a
little wool, or of moss and scraps of hay lined with rabbits' fur, willow,
or thistle down.

Position of nest.—Usually on a bed of chips in a hole in an old
willow tree, growing on the bank of a stream or river, near the ground,
or in some other suitable tree or stump in great woods, coppices
hedges, and swampy places.

Number of eggs.—8-12.

Time of nidification.—IV-V ; May.

The Marsh Tit affects low-lying marshy land dotted with
willows and alders, or in the vicinity of woods; also orchards
and gardens, in which, like the other members of its family,
it does good service in the destruction of insects ; though
occasionally seen in all the counties of England and most
of those of Wales, it is certainly far less numerous in some
counties than in others.

The call-note of this Tit is said to be harsh, and to
sound like the syllables " peh " " peh " harshly pronounced,
but the spring notes of the cock are varied, gay, and more
musical (Yarrell's Brit. Birds).

Mr. Harting has pointed out ('Zoologist,' 1867) that the
Marsh Tit differs from the Coal Tit in the absence of a
white spot on the nape of the neck, a character which is
always present in the latter species.

CRESTED TIT.

LOPHOPHANES CRISTATUS, *Linn.*

Pl. X., fig. 25.

Geogr. distr.—Found from the Mediterranean into Northern Scandinavia and through Europe from the extreme west, probably as far east as the Ural Mountains; in Great Britain it is chiefly confined to Scotland and the north of England, though occasionally found in the south and in Ireland.

Food.—Spiders, insects, seeds, and berries.

Nest.—A loosely-built structure of green moss lined with wool or fur of the mountain hare and feathers; sometimes more compact, formed like that of the Wren.

Position of nest.—In deserted Squirrels' or Crows' nests, hollow pine and oak trees, rotten stems of firs from twelve to fourteen feet high, in mountain ash trees by the roadside, or in juniper bushes.

Number of eggs.—4·6; usually 5.

Time of nidification.—IV·V.

This species frequents dense and old growths of pine and fir forest, being resident in a few of the oldest forests of Scotland in the counties of Ross, Inverness, Perth, Elgin, Banff, and possibly Aberdeen.

The nest, when placed in bushes, is stated to be formed like that of a Wren with a hole at the side.

The note of the Crested Tit is described as somewhat resembling that of the Coal Tit, but with a peculiar shake at the finish.

LONG-TAILED TIT.

ACREDULA ROSEA, *Blyth.*

Pl. X., fig. 27.

Geogr. distr.—Apparently restricted to Great Britain; a constant resident; common in certain districts.

Food.—Insects and larvæ.

Nest.—Usually a large and somewhat pear-shaped structure, the small end being upwards and pierced in front with a small entrance hole; but sometimes more barrel-shaped *; formed of mosses and fragments of grey and white lichens firmly compacted with cobweb and sometimes wool, feathers, and a few beech leaves; warmly lined with feathers.

Position of nest.—In stunted bushes, hawthorn hedges, box and other trees in groves, plantations, shrubberies, or gardens near woods.

Number of eggs.—8-10.

Time of nidification.—IV-V; May (beginning of month).

This is one of our most attractive birds, and its nest is surpassed in beauty only by that of the Chaffinch; it is a very conspicuous object in a hawthorn hedge, and therefore in such a position rarely escapes the eye of that most destructive animal the Clodhopper, who ruthlessly tears it to pieces, and, placing the eggs on the ground, tries to see how many he can smash at one jump, or with one shot of a stone at a yard's distance.

The flight of this Tit is rapid, and its cry is shrill; it leaves its eggs at once as one approaches the nest, but appears to be much attached to its young, which remain with the parent birds until the return of the nesting-season.

From observations recorded by Mr. Weir, the construction of the nest occupies both cock and hen (working alternately) for twelve days, and when one considers the labour expended in decorating its exterior, it is not at all surprising that so much time should be occupied in perfecting it; it varies much in size.

Although it is stated that as many as sixteen eggs are sometimes deposited in a nest, that number is hardly likely to be laid by one hen; in my experience ten is the usual number.

---

* These two forms of nest have probably earned for this bird the names of Bottle Tit and Barrel Tit.

Family **PANURIDÆ.**

BEARDED REEDLING (OR TIT).

PANURUS BIARMICUS, Linn.

Pl. X., fig. 26.

Geogr. distr.—Throughout Europe, in nearly all suitable localities; local in Great Britain, but resident.

Food.—Seeds of reeds, insects, small snails.

Nest.—Open and cup-shaped; formed of dead leaves of sedges, reeds, and grasses, interwoven with cobweb, and lined with the top of the reed.

Position of nest.—Near the gruund, in tufts of coarse grass or rushes growing in the fens, on the margins of dikes, amongst broken-down reeds, or on the edge of a mass of water-plants.

Number of eggs.—5-7.

Time of nidification.—IV-V; end of April or May.

The Bearded Tit is seldom found far from the dense reedy fens and water which it loves; formerly it was not uncommon on the Norfolk Broads; but now it is much rarer, though still occasionally found. In the spring of 1885 I noticed one or two unfinished nests, evidently of this species, amongst the reeds on the Ormesby Broads, but, unfortunately, I was too early to obtain the eggs. I have also seen a nest, but without eggs, at Kemsley, on the Kentish Coast. The latter was placed upon a small island of partly decayed reeds (trodden down by Moorhens and Ducks) near the centre of a large pond. It is therefore clear that the county of Kent is not yet wholly deserted as a breeding-place by this species, though Hickling and Horsey Broads in Norfolk are still probably its favourite resort.

Like the True Tits, the Bearded Reedling is addicted to gymnastic exercises, often hanging head downwards to feed, and running up and down the reed stems with ease; upon the ground, however, its mode of progression is awkward, the head being lowered, and the action of the bird a waddle rather than a walk.

Family MOTACILLIDÆ.

PIED WAGTAIL.

Motacilla lugubris, *Temm.*

Pl. X., figs. 29-31.

Geogr. distr.—Western Europe, southward to the western portion of N. Africa; common and generally distributed in Great Britain.

Food.—Larvæ, insects, Mollusca, and small fish.

Nest.—Formed of dry grass-stalks and rootlets or moss; thickly lined with wool or feathers and hair.

Position of nest.—In ivy on the top of a low wall, in a faggot-stack or a hole in the sloping grassy bank of a deserted chalk pit, or roadside bank; a deserted Sand Martin's hole, a heap of stones, or a hole in a wall.

Number of eggs.—4-6.

Time of nidification.—IV-VI; end of May.

I have known the Yellow Bunting, the Pied Wagtail, and the Robin to build in the same hole in the bank of a chalk-pit in consecutive years; as is usually the case with holes commonly selected for nesting purposes, this cavity and one or two others in the same pit were clearly visible from the road, distant from them about eighty feet or more; indeed, as a matter of fact, very few nests are so well concealed as to escape the practised eye of a naturalist after a few years' training, although in his first season or two of birds' nesting, his attention may only be arrested by the sudden flight of the parent bird from its habitation.

The nest varies in size and strength, some specimens being very loosely and carelessly constructed, whilst others are solid, firmly felted, and like some formed by the Greenfinch, but with more nearly the lining of a Redpoll's nest; the materials, however, vary but little: I have every stage of perfection before me, and, had I not taken them all myself, I should have been inclined to believe that the eggs (which are almost identical in all) had been placed in nests formed by other birds; they much resemble in colouring some of the lighter varieties of eggs of the House Sparrow, but are better formed, being less elongated and more pointed at the smaller end.

F

WHITE WAGTAIL.

MOTACILLA ALBA, *Linn.*

Pl. X., fig. 32.

Geogr. distr.—Widely extended throughout the Palæarctic Region; rare in Great Britain, but is known to have nested in the Isle of Wight and on the South Coast.

Food.—Insects, slugs, and other small Mollusca, and small fish.

Nest.—Formed of grass bents, rootlets, and moss, lined with hairs, and occasionally a few feathers.

Position of nest.—In a hole in the ground, a faggot-stack, stone heap, or wall.

Number of eggs.—5-7.

Time of nidification.—IV-VI; rarely as late as October.

The habits of this bird are very similar to those of the Pied Wagtail; though it has been stated that the White Wagtail does not follow the plough like our common species,* it is known to occur in Cornwall and N. Devon in spring. Mr. Broderick also found a pair nesting in a wall near Ilfracombe, and it is considered quite possible that it may breed in Kent, where the bird appears yearly; indeed, as both nest and eggs resemble those of the commoner bird, it is not impossible that they may have been taken for that species, and so not recorded: the birds, however, differ considerably, the head and neck being at first sight very similar †; but the back, scapulars, rump, and upper tail-coverts being pale ash-grey; the primaries, secondaries, tertials, and wing-coverts, brownish black; as a whole, there-fore, this bird is much paler in colouring, and ought to be readily distinguished from the Pied Wagtail; if it sits as closely as I have known the latter species to do, it might even be caught upon the nest without much difficulty, and in that case there could be no question as to the species in the mind of its captor.

---

* Until more well-authenticated instances of the discovery of the nest have been recorded, it appears to me to be premature to state anything definitely as to its *nesting* habits in this country.

† Mr. Harting, however, gives the following distinctive characters:— "The head covered with a distinct hood of pure black, perfectly defined and not mixing either with the grey of the back or the white of the forehead; the white on the cheeks and sides of neck completely separates the black of the head from that of the throat and breast, there being no black before the shoulders; the sides are also much lighter, and the tail somewhat longer. In the female there is no mixture of black on the back and nape, which there is in the female Pied Wagtail."

GREY WAGTAIL.
MOTACILLA MELANOPS, *Pallas.*
Pl. X., figs. 34, 35.

Geogr. distr.—Europe as far northward as N. Germany and Great Britain; Asia as far east as Japan and as far south as Java; partially resident, but local in Britain, breeding chiefly in the N. of England and Scotland.

Food.—Insects and Mollusca.

Nest.—Neatly constructed of rootlets or moss, and a few dry grass-stalks; thickly lined with cow's hair, usually of a white colour.

Position of nest.—Concealed by herbage upon banks under over-hanging rocks, or in forks of trees near a river or stream; also in fallow ground.

Number of eggs.—5.

Time of nidification.—IV-VI; commencing towards the end of April.

This species has bred in Cornwall, Devonshire, Somerset-shire, Dorsetshire, Wiltshire, Hampshire, Kent, Derbyshire, Cumberland, and Westmoreland; but it is commoner in the Northern Counties; the fact of its occurrence in Kent, though already recorded, is confirmed by a nest which I obtained at Kemsley, in Kent, on the 16th May, 1885; it was built on the ground behind a clod of earth in a fallow field, and was discovered by a boy whilst ploughing: this boy took me to the spot, and the bird flew off the nest as we approached. The nest is formed of root fibre mixed with coarse grasses, cow- and horse-hair, and is lined with black horse-hair, white cow-hair, and wool; it originally contained four eggs of the ordinary type, but one was broken by the fall of a piece of earth into the nest as it was being removed; they are slightly larger than the eggs of the Pied Wagtail, and are rather closely mottled with pale yellowish brown, which gives them a stone-grey tint.

The song of the Grey Wagtail is like that of the Meadow Pipit.

(BLUE HEADED) YELLOW WAGTAIL.

MOTACILLA FLAVA, *Temm.*

Pl. X., fig. 33.

Geogr. distr.—Central Europe and Asia during the breeding season; southern Europe and Asia or N. Africa during seasons of passage ; in Great Britain it has occurred in Cornwall, Devon, Somerset, Sussex, Kent, Essex, Suffolk, and Norfolk, chiefly from April to June.

Food.—Insects and Mollusca.

Nest.—Somewhat loosely constructed of fine rootlets, grass, straws, and bents, sometimes mixed with moss; lined with horse-hair, wool, or fine bents, occasionally with wool mixed with a few downy feathers.

Position of nest.—On the ground under a tussock, or in the bank of an old dry ditch amongst rather dense herbage.

Number of eggs.—4-5 ; rarely 6.

Time of nidification.—V-VI ; usually beginning of June.

About the first week of June, 1885, my friend Mr. William Drake forwarded to me a nest and eggs found by one of his boys amongst the grass in the saltings by the creek at Kemsley, near Sheppy, in Kent; the bird was known to the boy by the popular name of "Whip Jack," and was described as having a "black head with white ring," a near enough description to identify the species with the blue-headed Wagtail.* The nest is slighter than those of the three other species which I obtained at Kemsley in May ; it is formed of fine fibrous roots and coarse grasses, and is thickly lined with black horse-hair ; it contains six eggs of a pale, yellowish-brown colour, rather yellower than those of the Yellow Wagtail (*M. raii*), and not unlike some eggs of the Sedge Warbler, excepting that they are slightly larger and have a more polished surface.

The saltings at Kemsley are spongy and marshy, for the most part covered with wiry grass tussocks.

---

* I may mention that I have seen this species of Wagtail within a mile or two of Kemsley, and at the same time of year; it differs noticeably from the Yellow Wagtail in its grey instead of pale olive head, the white instead of yellow line over the eye, and the white instead of yellow chin and throat ; the tail feathers also differ.

YELLOW (or RAY'S) WAGTAIL.

MOTACILLA RAII, *Bonap.*

Pl. XL, fig. 1.

Geogr. distr.—Widely extended in Europe, but only common in the western countries; Western Asia and Africa; tolerably common in Great Britain, arriving in April.

Food.—Larvæ, insects, wood-lice, Mollusca, and small fish.

Nest.—Formed of dried grasses and rootlets, lined with finer rootlets, hair, bents, and sometimes with green moss, rabbits' down, or sheep's wool.

Position of nest.—On the ground, usually under shelter of a tussock or clod, and always fairly well concealed amongst herbage.

Number of eggs.—4-6.

Time of nidification.—V.

This species frequents meadows, old brick-fields, and arable land.

A nest of this bird was brought to me on the 16th May, 1885, by Mr. W. R. Salter (a cousin of my friend Mr. Drake) who found it at Kemsley, in Kent, in the bank of a pit partly filled with water, amongst rank herbage. The nest is large and well formed of coarse grass, moss, and root fibre, and neatly lined with wool covered with fine fibre and black and white hair ; the eggs are six in number, and are of a pale whitish stone-colour without any perceptible mottling ; they are, in fact, similar in colouring to the palest varieties of eggs of the common Partridge.

This Yellow Wagtail is by no means an uncommon species in the neighbourhood of Sheppy ; I have seen it every year either in the brick-fields of Murston or Kemsley, where it doubtless breeds regularly ; I have also seen the blue-headed Wagtail at Murston once or twice during the last fourteen or fifteen years, but there is no doubt that it is very rare in Kent, or I should have seen it more frequently ; for there is no possibility of mistaking one species for the other.

TREE PIPIT.

ANTHUS TRIVIALIS, Linn.

Pl. XL, figs. 3-5.

Geogr. distr.—North and Central Europe and Asia; migrating southwards about October to South Europe, Asia, and N. Africa; in Great Britain it is somewhat local; arriving in April and leaving in September or October.

Food.—Larvæ, insects, and slugs.

Nest.—Formed of dried grass and bents mixed with moss, somewhat finer towards the inside, and lined with a few coarse black hairs, twisted round as in the nest of the Whitethroat.

Position of nest.—In a hole in the ground, or in a bank, more or less concealed amongst the grass; in meadows surrounded by groves and shrubberies; in railway or roadside banks where the grass is tolerably long.

Number of eggs.—4-6; rarely more than 5.

Time of nidification.—V-VIII.

The Tree Pipit has been known to reach our shores as early as February, and to leave them as late as November; it breeds in most wooded and cultivated districts in Great Britain, but rarely in open, unenclosed country; it is, therefore, comparatively rare in West Cornwall, and not very numerous in Wales, whilst its occurrence in Ireland has not been satisfactorily established (Newton in Yarrell, 4th ed.).

Though the nest is usually more or less sheltered, one in my collection (containing a Cuckoo's egg) was discovered by Mr. Oliver Janson and pointed out to me, in the centre of an enclosed meadow of rank grass surrounded by plantations and garden shrubberies; his attention was drawn to it by seeing the bird fly off, otherwise it would have been as difficult to discover as the nest of the Sky Lark.

Probably few eggs offer more remarkable and unexpected variations than those of the Tree Pipit, as will, indeed, be seen by those which I have figured from specimens in my own series; every modification also occurs between the red variety and a form resembling the egg of the Meadow Pipit figured on my plate.

MEADOW PIPIT.
ANTHUS PRATENSIS, *Linn.*

Pl. XI.; fig. 6.

Geogr. distr.—Europe generally; in Asia it is recorded from Palestine, Trebizond, Teflis, Indersk, Persepolis, Western India, and Siam; N. Africa in winter, being common on both sides of the Straits of Gibraltar from October to the end of March, and found in Iceland from the end of April to September; common, generally distributed, and resident, in Great Britain.

Food.—Insects in all stages, worms, slugs, &c.

Nest.—Large; formed of grass stems, finer towards the inside, and mixed with rootlets and horse-hair.

Position of nest.—On the ground in meadows, or on the borders of heathy moors; sometimes concealed under a tuft of grass.

Number of eggs.—4-6; usually 5.

Time of nidification.—V-VII.

This species chiefly frequents the less cultivated parts of the country, whether lying high or low, dry or wet. According to Hewitson, in the choice of a nesting place it "is more partial to the margins of heathy moors" than to meadows and pastures, and he states that it is very abundant on those which border the lake district of Cumberland and Westmoreland; "in many of the northern parts of the country it is perhaps the most abundant of small birds, if, indeed, it be not the only one visible in the wilder and more desolate spots;" "to the eastern and southern counties of England it is a regular autumn visitor." (4th ed. Yarrell).

The eggs vary much less than those of the Tree Pipit, my figure 4 of the latter and figure 6 representing very fairly the ordinary amount of modification which one meets with; I have, however, a paler variety, the broad end of which is scribbled over with black lines like those of the Yellow Bunting.

This bird is smaller than the Tree Pipit, and has a longer hind claw to the foot.

ROCK PIPIT.

ANTHUS OBSCURUS, *Lath.*

Pl. XI., fig. 7.

Geogr. distr.—Northern Europe during the summer; Central and Southern Europe in the winter; in Great Britain it is common round the sea coast, being resident in Northumberland and Durham.

Food.—Small seeds, insects in all stages, Mollusca, and small Crustacea.

Nest.—Formed of fine dry grass-stalks, or sometimes sea-weeds, lined with finer grasses and a few black horse-hairs.

Position of nest.—On the ground on the slope of a grassy bank, under a large stone, or upon ledges of rock at no great height above the sea.

Number of eggs.—4-5.

Time of nidification.—IV-V; May.

This Pipit is confined to the cliffs, rocks, and lowlands close to the sea, where it is generally distributed; when disturbed it behaves somewhat like a Wagtail, flying from point to point, and, as it settles, agitating its body and tail with a vibratory motion.

The eggs vary in much the same manner as those of the Sky Lark, and, in a series collected by Mr. T. Copeland, at Uist, in May, 1884, and forwarded for examination to Mr. Harting, were several clutches which in all respects resembled eggs of that bird.[*]

---

[*] Subsequently a clutch of five eggs was kindly sent to me by Mr. Copeland.

FAMILY ALAUDIDÆ.

SKY LARK.

ALAUDA ARVENSIS, Linn.

Pl. XL, figs. 8-15.

Geogr. distr.—Palæarctic Region generally; common everywhere in Great Britain, and resident.

Food.—Insects, snails, worms, seeds, and tender grasses.

Nest.—Loosely constructed of dried bents and grasses, and lined with fragments of finer grass-stalks.

Position of nest.—On the ground in fields of clover, lucerne, cereals, beans, &c.; also in depressions in meadows, cemeteries, moor-land, &c.

Number of eggs.—4-5.

Time of nidification.—IV-VI; end of May.

The eggs lie in the nest with the small ends towards the middle; only four are said to be deposited at the first sitting; I have nests taken both in May and June containing only that number. The mother bird displays considerable persistency in attending to her young, and is very careful in concealing her nest as much as possible. A remarkable instance of this came under my notice some years ago, when visiting the Isle of Sheppy: happening to meet a shepherd, I asked him if he ever met with Larks' nests; he led me back about three yards, and pointed to a patch of water-weed at the edge of the path as a nest; I stooped down and found that the weed was regularly interlaced in the edge of the nest over which it fell, completely concealing it. I asked the shepherd how he had discovered a nest thus ingeniously hidden, and he replied by asking me two questions, "How far is it from any fresh water?" and "How could the weed get there by itself?" There can be no doubt that many nests are thus passed by the educated but unreflecting collector which do not escape the eye of those whom Nature alone has instructed; as a matter of fact, most Larks' nests are discovered by rustics: to the ordinary untrained eye they appear merely like a small hole containing four or five pebbles, and they therefore attract no particular attention; the hole itself is not so deep or so conspicuously lined as that of the Nightingale, otherwise it would doubtless be one of the most easy to find, for there can be no question as to its being abundant.

WOOD LARK.

ALAUDA ARBOREA, *Linn.*

Pl. XI., fig. 16.

Geogr. distr.—Throughout Central and Southern Europe ; eastward as far as the Ural Mountains ; local in England.

Food.—Insects, worms, slugs, seeds, green food.

Nest.—Formed of straws and moss, lined with fine bents, wool, and hair ; it is therefore more compact than that of the Sky Lark.

Position of nest.—Ou the ground, usually well concealed in the grass or in a clump of heather, in uncultivated places dotted with bushes and trees ; it has been found upon the stump of a felled oak.

Number of eggs.—4-5.

Time of nidification.—III-VII ; May.

Less numerous than the Skylark, this species is, nevertheless, not uncommon, and breeds freely in certain spots throughout the Midland and Southern Counties of England ; it seems, however, rarely to breed from Bedfordshire northwards, excepting in Derbyshire, S. Lancashire, the East Riding of Yorkshire and Westmoreland ; in Scotland it breeds in Stirlingshire, and in Ireland in the county of Dublin ; its choice of a nesting-place varies not a little, but it appears carefully to avoid land which is under cultivation. During the breeding season I have seen this species sitting and singing in a little roadside copse in the wooded country near Dover ; though generally considered a shy bird, it took no notice of the approach of myself and a friend, but continued to sing as we watched it. Unlike the Sky Lark, it does not rise perpendicularly and continue to hover and sing in one spot, but sometimes it soars to a great height and keeps flying in wide irregular circles, singing incessantly ; it is a little smaller than the common species, has a shorter tail, a conspicuous pale streak over the eye and ear-coverts, and is more distinctly marked on the breast ; its eggs are generally paler, broader, and smaller, though some varieties differ chiefly in their slightly inferior size.

Family FRINGILLIDÆ.

Sub-family EMBERIZINÆ.

CORN or "COMMON" BUNTING.

Emberiza miliaria, Linn.

Pl. XL, figs. 17, 18.

Geogr. distr.—Europe generally; eastward to Turkestan and Central Asia; generally distributed, but somewhat local, in Great Britain.

Food.—Insects, seeds, grain, peas, and beans.

Nest.—Somewhat loosely constructed (though less so than that of the Yellow Bunting), formed externally of sticks, straw, or coarse bents, and dry, broad grasses; this outer wall is, however, very thin, and almost immediately gives place to a tolerably densely woven texture of slender grass-stems and rootlets, lined with horse-hair; the cavity is deeper than in most nests of *E. citrinella*.

Position of nest.—In a depression in the ground, in wild patches in flat corn-growing districts; or in low bramble, sprouting hawthorn tufts, tufts of the pink Ragged Robin, grass, or other low-growing herbage, close to the ground.

Number of eggs.—4-6; usually 5.

Time of nidification.—V-VII; June.

According to some writers, the outer wall of the nest is partly formed of moss; this may at times be the case, but I have not found it so with the nests that I have taken. The eggs are sometimes much more like large eggs of the Yellow Bunting than are the two varieties which I have figured; fig. 3 of my Plate XII, if a little less pink and larger, would very fairly represent a not unusual form.

This is a somewhat local, but, in the south of England, a comparatively common bird, though far less so than the Yellow Bunting; it chiefly occurs in cultivated districts; the nest, according to Seebohm, is generally on the ground in a little depression, but sometimes it is slightly above it; such as I have found have been in the latter situation, but as, according to Yarrell, it appears to be usually placed in a field of pease or red clover, grounds upon which I rarely trespass, the probability is that I have passed a dozen nests for every one I have taken.

The appearance of the Common Bunting is rather ungainly than attractive, and its note is harsh and inharmonious, and has been compared to the sound of breaking glass.

YELLOW BUNTING.

EMBERIZA CITRINELLA, *Linn.*

Pl. XII., figs. 3-10; and Pl. XXXVII., fig. 10.

Geogr. distr.—Western Palæarctic Region; scarce in S. W. Europe occurs everywhere in Great Britain.

Food.—Insects, seeds, corn, berries, fruits.

Nest.—Loosely constructed, sometimes less so than usual; formed of an outer lacing of coarse straws, withered grasses, sometimes a few twigs, and rarely an edging of withered chestnut-leaves; internally of fine withered grass bents, and a sprinkling of rootlets with a few hairs in the inner lining.

Position of nest.—Usually low down, less frequently at four to five feet from the ground: in hedges, bushes, low scrub; in holes in grassy banks by the roadside, or banks of gravel and chalk-pits.

Number of eggs.—4-5.

Time of nidification.—IV-VIII; May.

Hewitson mentions moss as a component part of the nest of this bird; I have examined every nest that I have come across for some years past, but have never yet discovered a particle of moss in any of them; I do not believe that the Buntings, as a rule, are partial to that material.

In the 'Zoologist' for December, 1883, I have noted a somewhat aberrant nest, "from which the coarse edging was entirely absent," and which I found with three fresh eggs, upon which the hen was sitting, at Box Hill, on the 12th August, 1882: eggs have been found as late as September.

The Yellow Hammer (or Bunting) is a very abundant species, and in Kent one meets with it in every ramble; its brilliant colouring, unsurpassed by that of a canary, and its funny little never-varying song, make it a general favourite; in partially-cleared waste ground where there is short scrub of about a year's growth, its nest abounds, and the bird, sitting close until one is almost upon it, and then flying off with no little bustle, constantly calls one's attention to the existence of a nest where least expected; it is, however, so conspicuous an object that there is rarely any difficulty in discovering it.

If the nest of this species is taken, the bird will sometimes continue to lay in the same spot without forming another: this accounts for the Yellow Hammer having been seen sitting upon eggs laid on the bare ground, or, as I have seen, upon a mere platform of knotted ivy branches in a hedge.

CIRL BUNTING.

EMBERIZA CIRLUS, L*inn.*

Pl. XII., figs. 11, 12.

Geogr. distr.—Central and Southern Europe; eastward to Asia Minor; constant, but local, in Great Britain.

Food.—Insects, especially grasshoppers, grass seeds, oats, and berries.

Nest.—Tolerably compact, but with a loose outer wall of coarse grass, rough straws, vegetable fibre, &c.; within this a mass of inter-woven fibrous roots, a few fine bents, and a good lining of black horse-hair.

Position of nest.—Near the ground, in furze, bramble, or low bushes.

Number of eggs.—4-5.

Time of nidification.—V-VI; May.

I first took the nest of this bird—containing, however, only three eggs—on the 5th June, 1877, at Iwade (near Sheppy) in Kent; the bird was sitting, and I almost stepped upon her; the nest was in a low furze bush tangled with a Blackberry vine, and growing in a rough bit of scrub in a country lane. Two of these eggs are figured on my plate; compared with those of the Yellow Bunting, the eggs of *E. cirlus* are, as a rule, shorter, broader, whiter, less marked with scribbled lines, and more so with black spots and splashes.

According to Col. Montagu, whose description of the nest seems to have been handed down from one oologist to another, a little moss is intermixed with the other materials. Mr. Seebohm also describes a little in the lining of one in his collection; no doubt this is sometimes the case, just as it is absent from some nests of the Greenfinch, but hitherto I have not seen it.

The Cirl Bunting probably breeds every year in Kent, but not in great numbers; I have once or twice noticed the egg in small private collections, and have myself taken it about three times in twelve years; it is recorded as having bred in all the southern counties from Cornwall to Sussex, in Surrey, Middlesex, Hereford, Buckingham, Berkshire, Wiltshire, Gloucester, Worcester, and Warwick; it may always be known from the Yellow Bunting by its black throat; its song somewhat resembles that of the Lesser Whitethroat.

REED BUNTING.

EMBERIZA SCHŒNICLUS, *Linn.*

Pl. XII., figs. 1, 2.

Geogr. distr.—Extending widely throughout Europe, excepting in the Central and Southern districts; at the approach of winter migrating to N. Africa; in Asia as far east as Japan and as far south as India; in Great Britain it is generally distributed and resident.

Food.—Seeds, berries, and insects.

Nest.—Formed of dry grass and, according to most writers, moss; lined with finer grass and a few hairs of the feathery tops of reeds.

Position of nest.—On the ground in swampy land, amongst willows and osiers, or in a clump of long grass reeds or rushes and aquatic plants; rarely in a low bush; usually well concealed.

Number of eggs.—4-7; rarely more than 5.

Time of nidification.—IV-VII. May.

In Norfolk I have heard the name Water-Pheasant applied to this species; if one expected to find the Asiatic *Hydrophasianus* on the Broads, he would be slightly disappointed; yet the eggs of the Bunting are, in my opinion, prettier than the large, shining peg-top-like eggs of the true Water Pheasant.

In June, 1885, I received two nests of this species, one obtained by my friend Mr. William Drake on the saltings at Kemsley, near Sheppy, in Kent; this nest is constructed of fine bents and a few rough twigs, thickly and firmly lined with finer bents, fibre and hair; the eggs (five in number) are very decidedly and beautifully marked: the second nest was obtained at Downton, Salisbury, in May, by my friend Mr. Salter, a cousin of the above gentleman; it is a smaller and more coarsely constructed nest than that from Kent, being formed externally of coarse, reedy grass leaves, becoming finer and mixed with bents and horse-hair towards the centre; neither of these nests shows a trace of moss in its construction, nor have I hitherto seen it in nests of this bird; Mr. Seebohm, however, says that in some districts they are made externally of dry grass, moss, and withered leaves of rushes, and internally of fine grass and hairs.

The Reed Bunting rears two, if not three, broods in the year.

Sub-family *FRINGILLINÆ*.

CHAFFINCH.

Fringilla cœlebs, *Linn.*

Pl. XII., figs. 13-20 ; and Pl. XXXVII,. fig. 9.

Geogr. distr.—Throughout Europe from the extreme West to the Ural range and southwards into N. Africa; travels northward in summer and southward in winter; is extremely common in Great Britain, even to the north of Scotland.

Food.—Insects, seeds, berries, and fruits.

Nest.—Usually firmly and neatly constructed of moss, a few lichens, and some spiders' web, lined with fine roots and hair mixed with or completely covered with a layer of thistle-down.

Position of nest.—Most frequently in hawthorn hedges, especially when they enclose an orchard, but also often placed in forks of young fruit trees, or on moss-covered boughs of old apple trees.

Number of eggs.—4-6; more often the former.

Time of nidification.—IV-VIII ; end of May.

In the ' Zoologist,' vol. vii., p. 492, I have mentioned two nests which I took in 1882 and 1883, the first of which is "roughly constructed of roots and fibre, with very little adornment of moss and lichen, but a fair sprinkling of fine worsted" ; the lining is, as usual, of hair and thistle-down ; the second nest is chiefly remarkable from the fact that a conspicuous white feather is introduced into the outer wall.

The blue variety of the egg on Plate XXXVII. was one of a clutch of four taken on the 13th May, 1882, from a hawthorn hedge ; the nest was pointed out to me by my good friend Dr. John Grayling, of Sittingbourne, Kent ; both bird and nest were perfectly normal, and I was not a little surprised to see the close resemblance which the eggs bore to those of some varieties of the egg of the Bullfinch, which I had taken in previous years. A curiously imperfect and very shallow nest was taken by Mr. Oliver Janson from a hawthorn hedge at Albury in Hertfordshire, June, 1884 ; it contained four eggs, all of the blue type figured on Plate XII. (fig. 19), but entirely without markings, and with only a russet tinge at the larger end : this nest is now in my collection.

The Chaffinch is double-brooded.

GOLDFINCH.

CARDUELIS ELEGANS, *Steph.*

Pl. XII., figs. 1-3.

Geogr. distr.—Western Palæarctic Region, and as far eastward as Turkestan; generally distributed and partially resident in Great Britain, but rare in the north of Scotland.

Food.—Berries, seeds, flower-heads of weeds, young leaves, and insects in all stages,.

Nest.—Neatly constructed of moss and lichens interwoven with rootlets and wool; lined with thistle-down, short downy feathers, and horse-hair; it is small, cup-shaped and compact.

Position of nest.—Outer branches of chestnut, poplar, or fruit trees, or in a non-evergreen bush; Hewitson, however, mentions a nest found by him " at the top of a lofty laurel."

Number of eggs.—4-5; rarely 6.

Time of nidification.—V-VI.

This species frequents small groves, gardens or orchards in the vicinity of fields; I used formerly to take its nest occasionally in Kent; but during the last few years I have only once or twice seen the bird, and have not come across its nest; the zeal of professional bird-catchers probably does much to reduce the numbers of this much admired bird, but certainly less than the increasing cultivation of so-called waste land, and the consequent destruction of the weeds, upon the seeds, flower-heads, and leaves of which they feed.

The nest, though less compact than that of the Chaffinch, is, on the whole, not unlike it, and is placed in similar situations; the young are probably fed chiefly with small caterpillars.

Speaking of the places which the Goldfinch most affects, Mr. Seebohm says (Hist. Brit. Birds, vol. ii, p. 88) " It may often be seen in country orchards, and appears to have a partiality for the neighbourhood of houses, and near them it most commonly builds its nest."

The Goldfinch is often paired with the canary when kept in confinement; its song is not quite so loud, but almost as pleasing, as that of the Linnet.

SISKIN.
CHRYSOMITRIS SPINUS, *Linn.*
Pl. XIII., fig. 4.

Geogr. distr.—Europe generally, visiting N. Africa in winter; Asia as far eastward as China and Japan: in Great Britain it is chiefly confined to the northern counties of England and to Scotland.
Food.—Insects (*Aphides*, when young), seeds, and berries.
Nest.—Small, formed of spruce-twigs and bents, lined with cotton of *Salix* and a few feathers.
Position of nest.—Carefully concealed in bushes of spruce or juniper in forests.
Number of eggs.—4-5.
Time of nidification.—IV-VI; April.

Speaking of the habits of the Siskin in the breeding season, Mr. R. J. Ussher says, "In April and May, 1857, Siskins were unusually common at Cappagh, in the woods of fir, both on the low ground and on the hill side; in fact, the woods were continually ringing with the song of this bird. You might hear it as it flew over the wood uttering its peculiar cry, half chirp, half song; at one time flying straight forward, as if to some destination, then turning and making a circuit, as if it did not know its own mind, or as if it were loth to descend from its joyous flight, then again darting off in a new direction, whilst its notes would gradually die away. Its every tone and movement is full of animation and delight, as if it were beside itself with pleasure; this is particularly the case in the nesting-season, at which time I have seen the male flying slowly towards some topmost spray of a fir tree, pouring forth his delightful little warbling song, which very much resembles that of a Goldfinch, but is to my ears far sweeter. It very often sings when flying, but more frequently when perching on some fir-tree top; indeed the Siskin in spring seems more like a visitor from a happier world."

The Siskin is for the most part an autumn visitor to England, arriving in flocks towards the end of September, and remaining with us until April; it, has, however, been recorded as breeding in Kent, Surrey, Dorset, Sussex, Middlesex, Oxford, Gloucester, Denbigh, Bedford, Derby, York, Westmoreland and Durham. Two broods are reared in the year, the first eggs being laid in April, the second in June. The call-note is " *hootelee, hootelee.*"

LINNET.

LINOTA CANNABINA, *Linn.*

Pl. XIII., figs. 5-10.

Geogr. distr.—Widely extended throughout Europe; a winter visitant in N. Africa; common and generally distributed in Great Britain; rarer in Scotland than in England.

Food.—Seeds and insects.

Nest.—Compact and cup-shaped, rarely so large as that of the Greenfinch, the walls not often exceeding 1 to 1½ inches in width, and seldom with moss or, at any rate, with any quantity of it, amongst its materials; the following in my collection are a few which I noticed in the 'Zoologist' for December, 1883 :—1. Slightly built for the species, but the walls strengthened with coarse straws, evidently selected from a dunghill." 2. "Not unlike nest of Yellow Bunting; its construction, however, decidedly firmer." 3. "Untidy, loosely put together, and has blackish straggling roots projecting from the sides." 4. Unusually deep, formed of roots, fibre, and wool, with a few white hairs towards the interior.   5. "Very ragged, formed of coarse bleached roots, lined with fine fibre and wool."

Position of nest.—In hawthorn hedges, bushes, and low shrubs, in forks of young trees, in furze, heather, and brambles near the ground, and occasionally even on the ground.

Number of eggs.—4-6; most frequently 5.

Time of nidification.—IV-VI; May.

A double-brooded species, and apparently with a very strong parental instinct; it frequently attracts one to look at its nest (though only the beginner will probably care to take it) by giving the alarm as one approaches its place of concealment; when sitting it is not very easily driven from the eggs, only leaving them at the last moment: if one, or even two, eggs are abstracted it will also not invariably desert the remainder; and this, though few non-collectors will believe it, is an act of wonderful forbearance on the part of almost any bird excepting a Tit.   The nest is abundant in most situations, but especially so in furze bushes, on the top of a loose, low growing whitethorn hedge, or in the forks of saplings in dense thickets.

BRAMBLING.

FRINGILLA MONTIFRINGILLA, Linn.

Pl. XII., fig. 21.

Geogr. distr.—Throughout the Palæarctic region; a winter visitant in Great Britain; it has been known to breed in Scotland, and is well known to remain with us as late as April.

Food.—Seeds, grain, beech-mast, berries and insects.

Nest.—Much larger and deeper than that of the Chaffinch; formed of green mosses and fine bents compacted with cobweb, ornamented with pieces of white lichen and shreds of the fine white paper-like outer bark of the birch; lined internally with fine wool and feathers.

Position of nest.—From ten to thirty feet from the ground, in forests of birch trees.

Number of eggs.—5-8.

Time of nidification.—V-VII; June.

At least one instance of this bird having bred in confinement has been recorded. Its song begins with a low warbling, succeeded by a harsher and more protracted note. Montagu says, "We have not been able to discover that this bird has ever bred with us, but they are frequently seen in the winter, in large flocks, upon the coast of Kent and Sussex when the weather is severe, and have been so exhausted as to suffer themselves to be taken up. They are also found in the interior parts of the kingdom at that season, flying in company with Chaffinches and Yellowhammers."

In his 'Catalogue of Birds of the Dyke Road Museum at Brighton,' p. 126, Mr. E. T. Booth says, "In the summer of 1866, while fishing on the River Lyon, in Perthshire, I had occasion to climb a beech tree to release the line which had become entangled in the branches, and while so engaged a female Brambling was disturbed from her nest, containing three eggs, which was placed close to the stem of the tree. As I was anxious to procure the young I left her, and, on again visiting the spot in about a fortnight the nest was empty, and, judging by its appearance, I should be of opinion that the young birds had been dragged out by a cat. This is the only instance I have ever known of the Brambling attempting to rear its young in Great Britain."*

Mr. Harting informs me that this bird once nested in Yorkshire, at Baldersly Park, near Thirsk. The nest, with five or six eggs, was taken by the Hon. Guy Dawnay, and forwarded to the Rev. J. C. Atkinson, who recorded the fact in the Field for the 23rd July, 1864, p. 52.

---

* I have to thank Mr. Harting for calling my attention to the above valuable note.

TWITE.

LINOTA FLAVIROSTRIS, L*inn.*

Pl. XIII., fig. 11.

Geogr. distr.—Europe generally, breeding in the north and wintering in the south ; in Great Britain it breeds in the Northern, Midland, and Western counties ; it is common in Wales.

Food.—Seeds, berries, and fruits.

Nest.—Formed of fibrous roots, plant-stalks or heath and grass externally : towards the inside of fine roots and feathers, and lined with wool, rabbits' fur, or thistle-down and hair.

Position of nest.—In trees or walls, amongst long heather, or upon the ground under strips of turf turned up by the plough.

Number of eggs.—5-6 ; usually 5.

Time of nidification.—V-VI ; May.

The Twite frequents moors and open heath-covered places. In habits it resembles the Linnet. It breeds not uncommonly "in the more hilly districts of Hereford, Salop, Stafford, Derby and Chester, as well as in North Wales and the Isle of Man, and on elevated moorlands in the higher glens, with increasing frequency northward from Lancashire and the West Riding of Yorkshire to Shetland, though in some districts it is rather scarce, and its stronghold in the West of Scotland is the Outer Hebrides. In Ireland it is found from north to south, and probably breeds in suitable localities throughout the island." —(Newton in Yarrell's ' British Birds,' vol. ii., p. 161.)

The Twite, though very like the Linnet, is a smaller bird, has no red on the crown or breast, and has a longer tail. Its song, though somewhat similar, is more monotonous and less powerful ; its call-note is, however, said to be shriller and somewhat to resemble the word " twite," whence its name has been derived.

To attempt, as has been done, to describe differences of marking by which the egg of the Twite may be distinguished from that of the Linnet, is useless, as the eggs of the latter bird are sometimes, and not unfrequently, wholly unmarked, whereas at other times they are as heavily speckled and spotted as the best marked eggs of the Greenfinch (*cf.* my figs. 8 and 10). The egg which I have figured was selected by Mr. Seebohm from his magnificent collection as a fairly typical form, but I have seen genuine Linnets' eggs very nearly resembling it.

LESSER REDPOLL (OR REDPOLE).
LINOTA RUFESCENS, *Vieill.*
Pl. XIII., fig. 12.

Geogr. distr.—Restricted during the breeding season to Great Britain, but proceeds later in the year to Western Europe. Breeds chiefly in the northern counties of England and in Scotland. Resident throughout the year in wooded districts.

Food.—Seeds of groundsel, thistle, &c., which it picks out whilst clinging to the plant-stems.

Nest.—Neatly and usually firmly constructed, small and elegant, formed of plant-stalks, roots, moss, and dry grass, with hair towards the interior, which is beautifully lined with pure white willow-down, wool, or occasionally with fine grasses and feathers.

Position of nest.—In low trees or bushes near water.

Number of eggs.—4-6.

Time of nidification.—V-VI.

I have twice taken the nest of this species from grass tussocks growing upon narrow paths through water and marsh land at Murston in Kent: in each case the nest contained six eggs; but those in the first nest which I discovered were somewhat incubated, so that one of the eggs was destroyed in blowing. In both cases I flushed the hen off the nest. I also have an aberrant nest with unusually pale eggs, taken from a bush in marshy ground at Kemsley. It is compact, but by no means firm, being formed almost wholly of wool, thinly covered externally with dry grasses and internally with hair. I have seen the cock bird during the breeding season (and I took pleasure in watching it for fully five minutes) hopping in and out of an open hawthorn fence at Detling, in Kent; so that there can be no question whatever that the species breeds in this county, though somewhat sparingly; indeed it has already been recorded as breeding there by Mr. Wharton (Zool. p. 8951). It also breeds—at any rate, occasionally—in Dorset, Hampshire, Middlesex, Oxford, Gloucestershire, Salop, Worcestershire, Warwick, Lincolnshire, Cambridge, Norfolk, Suffolk, Lincolnshire, Nottinghamshire, Derbyshire, Cheshire, and thence in all the counties northwards. The cock Redpoll, in breeding plumage, is a prettily-coloured bird, with his yellow bill, crimson crown, and rosy throat, chest, and sides of breast, and (if not often seen) is sure to attract one's attention.

TREE SPARROW.

PASSER MONTANUS, *Linn.*

Pl. XIII., figs. 13-16.

Geogr. distr.—Europe generally, N. Africa, Asia as far east as Japan: locally distributed and resident in Great Britain; it has been found in Middlesex, Surrey, Kent, and throughout the South of England.

Food.—Seeds and grains, berries, buds, green leaves, fruit.

Nest.—Roughly constructed of straw, hay, roots, wool, hair and feathers; thickly lined with feathers.

Position of nest.—In holes in trees, and occasionally in clefts of rock or old walls; but, in Europe, it objects, as a rule, to building in houses.*

Number of eggs.—4-7; usually 5.

Time of nidification.—IV-VIII; May.

A three-brooded species.

The young collector frequently supposes that the nest of the House Sparrow, when built in the fork of a tree, is necessarily that of the Tree Sparrow; the latter, however, builds only in holes, and, as the House Sparrow occasionally selects a hole in a tree for the same purpose, it is best to see the birds themselves in order to be sure of correctly identifying the eggs. The latter are smaller, and, as a rule, darker than those of the commoner species, although in both birds they vary considerably, as will be seen in my plates.

As would be expected from the inferior size of its eggs, this species is slightly smaller than the House Sparrow, *the head and nape chestnut-coloured*, with a spot behind the eye, and the chin black, the body above rufous-brown spotted with black, and with a greenish shade behind; sides of the neck, breast, and under parts dull white; the wing-coverts rufous, edged with black and crossed by *two white bars*, the greater coverts black, with rust-reddish edges; the quills blackish with rufous edges, *the tail rufous-brown, legs pale yellow.* Its chirrup also differs from that of *Passer domesticus*, being shriller and having more claim to be called a song, though only consisting of a repetition of its call-note.

I have taken the eggs of this bird both in Kent and Norfolk. The nest is usually placed in a hole in an old pollard willow.

---

* Seebohm says, " It is never seen in the towns (in our islands) but sometimes approaches the villages, where it associates with the House Sparrow."

COMMON SPARROW.

PASSER DOMESTICUS, *Linn.*

Pl. XIII., figs. 17-25; and Pl. XXXVII., fig. 5.

Geogr. distr.—Europe generally, excepting the extreme north; abundant and resident in Great Britain.

Food.—Seeds, grain, refuse food (both farinaceous and animal) insects.

Nest.—An enormous structure, usually carelessly put together, somewhat after the pattern of a Wren's nest; formed externally of a quantity of straw, hay or grass bents, sometimes mixed with rags, paper, string, and occasionally a quill-feather or two from a poultry yard; lined warmly with feathers, thin or crumpled paper, and other soft materials.

Position of nest.—In ivy on a wall, in holes in walls, chimneys, or occasionally trees, in holes bored by the Sand Martin: in thatches or under eaves; frequently in tall trees or hedges.

Number of eggs.—4-6.

Time of nidification.—III-VI*; May.

This is the most pertinacious and impudent, as well as the most destructive and least useful of our British birds; it appears to breed on and off for nearly the whole year, cases having been recorded of its rearing a brood in the depth of winter. Not only does it annoy one by littering the whole place with stubble and dust-hole refuse (which it heaps up under the impression that it is building a nest), with smashed eggs and writhing half-dead squabs, which, having fallen out with one another, have ended by falling out of the nest; not only does it wake up one at daybreak (however early that may be) by its senseless chirps and twitterings, but it drives from the premises all more attractive birds which chance to be weaker than itself. For this service it pays itself by plucking to pieces every spring flower in the garden, eating every seed or seedling which is not protected, reducing carnations to shapeless stumps, scuffling about on the ground so as either to litter the path with earth or the flower-beds with gravel stones, occasionally picking up some fine juicy spider or soft green caterpillar as a pretence of doing good; in short, it is amongst birds what the *Aphis* is amongst insects.

---

* This species is, however, very irregular, and has been known to breed in mid-winter as mentioned hereafter.

GREENFINCH.

LIGURINUS CHLORIS, Linn.

Pl. XIV., figs. 1-8.

Geogr. distr.—Throughout Europe, excepting the extreme North; common on the western side of North Africa: common and resident in the British Isles.

Food.—Seeds, grain, fruit and insects.

Nest.—Rather bulky, the walls varying in thickness from one to two inches, very firmly constructed as a rule, though occasionally loosely put together: very variable as regards its materials, the following being some of the types:—1. An outer framework of rough twigs and coarse roots, the walls of fine roots and green moss, and the lining of fine reddish fibrous roots. 2. Of coarse, half-decayed straws, bents, and roots, thickly lined with fine root-fibre. 3. Of slender, withered grass-straws and a mass of greyish wool felted together and lined with a few black horse-hairs. 4. Of sticks, roots, and moss, externally; of wool, vegetable fibre, and less moss, inwardly, lined with a few black horse-hairs. 5. Of coarse, plaited roots externally; of finer roots, moss, slender white hairs, and a little wool, matted together inwardly, and lined with a few black horse-hairs. 6. Of green moss with a few twigs and roots, and thickly lined with cocoanut fibre and a few black hairs. 7. Loosely constructed of green moss and spiders' webs, with a few twigs; lined with vegetable fibre and a few black hairs. 8. Of twice the usual depth (like a nest within a nest), formed of green moss, wool, and fibrous roots in patches, which give it an extremely soft and variegated appearance, a few twigs outside, and a little hair in the lining.

Position of nest.—In hawthorn hedges and bushes: rarely in loose hedges; in furze, laurustinus or other bushes, in low trees or on stumps of branches of tall trees; rarely very far from the ground.

Number of eggs.—4-6; rarely less than 5.

Time of nidification.—IV-VI; May.

This is one of the commonest of our birds, its nest being most abundant in hawthorn hedges or clumps of tall furze bushes. The eggs are sometimes heavily blotched with large chocolate-brown splashes.

Mr. E. T. Booth, in his 'Notes on British Birds,' (Pt. VIII.) mentions a curious nest of the Greenfinch, nearly twice the usual size, placed at the height of about six feet in a privet bush, and having the foundation entirely composed of a large mass of the Common Stonecrop (*Sedum acre*) torn up from a rockery close at hand.*

A very interesting nest was obtained for me by my friend Mr. Salter; he found it on the 9th June, 1885, at Downton, in Salisbury, built at a height of eight feet from the ground

---

* For this note I am indebted to Mr. Harting.

upon twigs which had sprouted from the trunk of a tall oak tree by the side of the road. The nest is almost of the shape of a sabot, and was placed sideways against the trunk, the foot of the sabot being a solid shelving mass of rough roots, fine fibre and coarse moss, felted together with vegetable wool; a narrow wall of the same materials surrounds the cup, which is thickly lined with white wool, a few slender fibres, and a single white hair; the cup is unusually deep, and the nest altogether has not merely an abnormal but a very attractive appearance.

HAWFINCH.

COCCOTHRAUSTES VULGARIS, *Pall.*

Pl. XIV., fig. 9.

Geogr. distr.—Central and Southern Europe; ranges eastward to Japan and southward to N. Africa. Resident in Great Britain.

Food.—Seeds, berries, peas, fruit.

Nest.—Variable in structure, some nests being more compact than others; it is formed of twigs, roots, and dried plants, varied with pieces of grey lichen, lined with finer roots and hairs; it resembles the nest of the Bullfinch on a larger scale.

Position of nest.—In high thorns, hornbeam, holly, boughs of oak, fir, or fruit trees, growing in groves and orchards.

Number of eggs.—3-6; usually 4-5.

Time of nidification.—V-VI; May.

According to Doubleday and Hewitson, the nest of this species is "most commonly placed in an old scrubby whitethorn bush, often in a very exposed situation."

According to Newton, "it is in what are known as the Home Counties, Middlesex, Essex, Hertford, Buckingham, Berks, Surrey, and Kent, that the Hawfinch is most plentiful, and its abundance in the last is shown by the fact that in 1876 Lord Clifton knew of more than fifty nests at Cobham. Mr. Cecil Smith has reason to believe that it has bred in Somerset, and to the eastward of long. 2° W. it has been ascertained to breed in every county south of York, save Stafford, Leicester, and Lincoln, in all which, however, the discovery of its nest is probably only a matter of time."—(Hist. Brit. Birds, 4th ed., vol. ii., p. 101).

It is a singular thing, in face of the above fact that, though nearly the whole of my birds'-nesting has been done in Kent, between Herne Bay and Maidstone, I have never yet come across a single nest of the Hawfinch.

The song of this bird is rather poor, but not altogether unmusical; the bird itself, though of a somewhat thick-set plebeian build, is very handsome in its colouring.

The nest, which is somewhat saucer-shaped, is usually placed in an apple or pear tree in an orchard, or in an old whitethorn, frequently in an exposed situation. Seebohm says that it "does not often build in shrubberies, and its nest is somewhat rarely placed in evergreen trees; but it has been found amongst ivy."

Sub-family *LOXIINÆ*.
COMMON BULLFINCH.
PYRRHULA EUROPÆA, *Vieill.*
Pl. XIV., figs. 10-15.

Geogr. distr.—Resident in Central and Western Europe; its range extending as far north as the British Isles; generally distributed and resident in wooded districts throughout England, and probably (though less commonly) in Ireland; in Scotland from Invernesshire to the south of Wigtownshire.

Food.—When young, insects; when adult, seeds, buds, berries, and fruit,

Nest.—Shallow and saucer-shaped, loosely but firmly constructed, externally of thin, dry, interwoven twigs of birch, fir, &c.; lined with fine roots, bents, hairs, and rarely a few leaves.

Position of nest.—Generally in a low bush or tree in a wood, grove, garden, or thick hedgerow.

Number of eggs.—4-5.

Time of nidification.—IV-VII; early in May.

Formerly I used to take the nest of this bird pretty commonly in Kent, but of late years the industry of bird-catchers has greatly reduced their numbers. The nest is at first sight not unlike that of a Wood Pigeon, though, of course, much more diminutive; but there is a greater distinction between the external framework and the lining, the latter being comparatively finer.

Speaking of the notes of this bird, Mr. Seebohm says, "The call-note of the Bullfinch to his mate is a full, rich, but low, piping whistle, very monotonous and plaintive, and sounds like *dyu dyu*; and his song, which is usually uttered in so low a tone as to be scarcely audible at a distance, as if he were fearful of being discovered in the act, is very pleasing and mellow. Usually it is warbled as he sits bolt upright, and every now and then jerks his wings and tail and turns his head from side to side, as if about to take wing."—(Hist. Brit. Birds, vol. ii., pp. 52, 53.)

The variety which I have represented at fig. 15 I have only once had the pleasure of taking; the nest was discovered and pointed out to me by a lady friend (Mrs. Homewood) in a bush in a wood at Stockbury, Kent; it contained two eggs which I still have, but my friend wished to keep the nest. The similarity of these eggs to those of the Greenfinch is somewhat remarkable.

COMMON CROSSBILL.
Loxia curvirostra, Linn.
Pl. XIV., fig. 16.

Geogr. distr.—Entire Palæarctic region ; an irregular visitor to Great Britain, but resident in some parts of Scotland.

Food.—Insects, seeds, berries, and fruits.

Nest.—Formed of long, broad grass or bents and plant-stalks, and coarse moss, felted together with wool or fine moss and lichens, and lined with wool or horse-hair ; the outside is supported upon a platform of twigs of larch-fir.

Position of nest.—In forks and branches of Scotch fir, occasionally not more than five feet from the ground, but usually at a considerable elevation.

Number of eggs.—2-3.

Time of nidification.—II-V ; March.

This species is believed to have bred in Devon, Somerset, Hampshire, Surrey, Sussex, Kent, Herts, Gloucestershire, Essex, Suffolk, Norfolk, Bedford, Leicester, Cumberland, Yorkshire, and Northumberland ; it has also bred in various counties of Scotland and Ireland.

" The attention of a passenger," says Newton, " is mostly drawn to the presence of a flock of Crossbills in one or other of two ways. He may notice the ground strewn with the fragments of enucleated cones, or—and this possibly the more often—he may hear a strange call-note, which has been syllabled *jip, jip, jip,* frequently repeated, and, on looking up, will find that it proceeds from birds that are ever and anon flying out from the branches of a tree, generally a conifer, and resettling upon it. Then he can stop to watch their actions carefully, for these birds are almost invariably tame, and admit of a very close approach, so much so, indeed, that instances are not uncommon in which they have been ensnared by a running noose affixed to the end of a long pole or fishing-rod and passed over their head, or, touched with a limed twig, adroitly applied by the same means, fall helpless victims to the ground." —(Hist. Brit. Birds, 4th ed., vol. ii, p. 195.)

Seebohm says that the female sits very close, and is fed on the nest by the male, and that the nest " is formed on the same model as that of the Bullfinch, an outside nest of sticks and an inside nest of soft materials, the latter rising somewhat higher than the former."

Family STURNIDÆ.

COMMON STARLING.

Sturnus vulgaris, Linn.

Pl. XIV., figs. 17-19.

Geogr. distr.—Throughout Europe generally; partially resident; Asia as far eastward as E. Siberia ; N. Africa and the Azores : resident and common in Great Britain.

Food.—Worms, slugs, insects, grain, fruits, birds' eggs.

Nest.—Formed roughly of dried grasses or chips of straw.

Position of nest.—In holes in walls, chimneys, ruins, rocks, caves, or trees; in deserted holes of the Sand Martin ; under thatches or in holes in the roofing of cottages and summer-houses.

Number of eggs.—4-6 ; rarely 3.

Time of nidification.—III-VI; May.

I have in my collection the top of a Starling's nest which I took out of the stove-pipe to a conservatory at Upchurch, in Kent; the pipe was bent almost at right angles to bring it up to the wall of the dwelling-house, above the roof of which it projected about a foot. The whole of the pipe from the angle to within about twelve inches of the top was filled with chips of straw, a whole straw being placed perpendicularly at the sides here and there, so that when the pipe was taken to pieces the nest was pushed out in the form of a cylinder about twelve feet in length. The whole of this material had been collected subsequently to the arrival of mild weather and the consequent disuse of the stove. The nest contained three eggs, tolerably hard set. If even these had been hatched, it is doubtful if the birds, when fledged, could ever have escaped from the pipe, which only measured about four inches in diameter ; indeed, it must have been extremely awkward for the parent birds to scramble out of and into this cylindrical dwelling.

To the florist the Starling is a friend, but to the fruit grower an object of aversion; yet for the good that this bird does methinks he might be spared a few poor cherries, and if one tree were reserved for him and the others covered with nets, surely no great mischief would ensue.

As the Starling robs the Rook, so is it in turn mobbed and robbed by Sparrows.

FAMILY CORVIDÆ.

RED-BILLED CHOUGH.

PYRRHOCORAX GRACULUS, Linn.

Pl. XIV., fig. 20.

Geogr. distr.—Occurs locally in Central and Southern Europe; eastward as far as China, and in N. Africa: in Great Britain it frequents the cliffs of the southern and western counties; resident.

Food.—Insects, Mollusca and grain; possibly also reptiles, small birds, and carrion.

Nest.—Externally formed of large and old sticks, lined with roots, wool, and hair.

Position of nest.—In crevices of rocks, sometimes out of reach from the opening.

Number of eggs.—4-5.

Time of nidification.—V.

I have seen this bird flying pretty commonly about the cliffs at Clifton, and on the coast of Devon, but have not hitherto had the pleasure of taking its nest; its note, though somewhat like that of the Jackdaw, is more shrill. When kept in confinement it exhibits similar mischievous propensities, pugnacity, and preference for certain persons, by whom, also, it takes pleasure in being caressed.*

"Years ago," says Seebohm, "the bird bred on almost all the suitable cliffs of the south coast; but at the present day most of its breeding stations are deserted. It still breeds in Cornwall, the north of Devon, on Lundy Island, and at many places on the Welsh coast, in Glamorgan, Pembroke, Anglesey, Flint, Denbigh, and possibly on the rocks of the Calf of Man." "In Scotland . . . . the great stronghold of the Chough is in the Island of Islay. On the west coast of Skye (which locality now appears to be its northern limit in our islands), in Wigtonshire and Kircudbrightshire a few pairs are still known to breed." In Ireland . . . . it is still common on the coasts of county Kerry, and I specially remember its abundance on the magnificent cliffs at Sybil Head, west of Dingle."—(Hist. Brit. Birds, vol. i., pp. 576, 577.)

---

* Montagu's 'Ornithological Dictionary,' ed. 1866.

COMMON RAVEN.
Corvus corax, Linn.
Pl. XV., figs. 1, 2.

Geogr. distr.—Throughout the Palæarctic and Nearctic regions, most numerous in the northern portions; also in Great Britain it is fairly common, though much less so than formerly.

Food.—Grain, berries, fruit, insects, reptiles, eggs, young poultry and game, rats, moles, and carrion.

Nest.—A bulky structure of rough sticks, lined with roots, wool, hair, and other soft substances.

Position of nest.—In forks of the larger branches of tall trees, in trees overhanging precipices, or in old ruins or inaccessible fissures in rocks.

Number of eggs.—4-6; frequently 5.

Time of nidification.—II-III; March.

The cry of this bird, when on the wing, is so much like that of a Drake, excepting for the want of repetition,* that a short time since I was completely deceived by it. The sounds uttered by the *Corvidæ* are, indeed, all different, and not one of them is at all like "*caw*," any more than the "*quarck*" of the *Anatidæ* is like "*quack*;" indeed the true sound is "*Whark-whark-whark-whark-whark*," and therefore the *q* is nothing more than the terminal *k* of a preceding utterance.

Seebohm says that "one of the best places in the British Islands to study the Raven's habits is the Western Isles of Scotland." "It is omnivorous, and will take almost everything in its power. Like the Hawks, it catches small birds and quadrupeds, kills a weakly lamb or fawn, and carries off the eggs of poultry and game should it happen to discover them; and it will never refuse to make a meal on carrion of any kind. Most animal substances are eaten—every creature which the sea casts up on the beach, from a dead whale to a mollusk; and it may sometimes be seen searching the pastures for moles, worms, and even insects. In autumn the Raven will also feed largely on grain."—(Hist. Brit. Birds, vol. i. pp. 533 and 535).

Owing to its tendency to attack weakly lambs, this bird is detested by shepherds as much as it is by farmers and rearers of poultry.

---

*It has been described by the words "*pruk, pruk*;" but the note sounds to me more like "*whurk*;" Seebohm gives it as "*cruck*."

Mr. Harting informs me that in March, 1864, a pair of tame Ravens which had the run of a garden belonging to Mr. Winterbottom, of Cheltenham, built a nest in a box in a shed about six feet from the ground. The nest was built of sticks, old fern-leaves, and the stalks of dead wallflowers, and was lined with dead leaves and tufts of grass. On March 4th two eggs were found in the nest, and the following day a third was laid; but the hen bird did not sit well, perhaps because too much disturbed by visitors, and the eggs were not hatched.

During the last winter (1885-6) I, on three occasions, saw a Raven fly over or close by my Surrey garden. I am informed by a friend that in the spring of 1885 a pair nested in the neighbourhood.

CARRION CROW.
CORVUS CORONE, *Linn.*
Pl. XV., figs. 5-6.

Geogr. distr.—Entire Palæarctic region, from the extreme west to Japan and China ; common in the southern counties of England, and resident.

Food.—Grubs, worms, Mollusca, refuse fish, reptiles, carrion, young birds, poultry, and game.

Nest.—Bulky ; formed of sticks and twigs, plastered with earth and lined with wool, hair, moss, or other soft materials.

Position of nest.—Generally in the forked branch of a tree, some-times that of the fir ; or among rocks.

Number of eggs.—4-6.

Time of nidification.—IV-V.

This species is solitary in its habits, and is seldom seen in this country in numbers excepting when attracted by carrion, or when roosting, in the winter. It is the terror of the shepherd, as it will tear out and devour the eyes and tongues of sickly sheep and lambs ; it is therefore fortu-nate that even two nests are rarely seen near together.

Seebohm remarks that the breeding season of the Carrion Crow is somewhat late ; and in this respect it differs con-siderably from the Raven, or even the Rook, approaching most closely the Jackdaw. The Raven's eggs are said to be often laid in February, the Rook's in March ; but the Carrion Crow seldom commences nesting duties until the latter end of April or beginning of May. It is very probable that this bird pairs for life ; and each season the old nest will be visited and used again, provided the owners are not molested.

Some years since, whilst at Iwade, near Sheppy, I met a boy with four eggs of this bird, which he had taken shortly before from a nest at the top of a tall elm tree close by. One of these eggs was no larger than that of a Turtle Dove, and, although the boy had already ruined all four by boring them with a pointed stick in order to get rid of the contents, I purchased them of him for the sake of this abnormal specimen. It was probably what a keeper of poultry would call a " cock's egg "—an egg without yelk ; such eggs of the Domestic Fowl are of common occurrence, and I have one or two as small as a Linnet's egg, but they seem to be of rarer occurrence with wild birds.

H

HOODED CROW.

CORVUS CORNIX, *Linn.*

Pl. XV., figs. 3, 4.

Geogr. distr.—Europe generally, though more common in the east than the west; in N. and Central Asia, and N. Africa; in Great Britain it is commoner in the northern than the southern counties.

Food.—Grubs, worms, Mollusca, carrion, reptiles, small birds, and grain.

Nest.—Similar to that of the Carrion Crow: bulky; formed of sticks or large branches of sea-weed, and lined with dry grass, wool, hair, and other soft materials.

Position of nest.—In trees, usually at a considerable height, or upon cliffs and isolated rocks.

Number of eggs.—5.

Time of nidification.—IV-V.

The eggs exhibit the same variations as the Carrion Crow, and therefore are not distinguishable from them; many eggs labelled in British collections as of this bird are doubtless those of the Carrion Crow, as *C. cornix,* though it occasionally breeds in England and Wales, is only regularly resident in Scotland and Ireland.

The Hooded, or, as it is sometimes called, the Grey Crow, is, in the British Islands, chiefly confined to moorland districts; in England it is an autumnal immigrant, appearing regularly in the autumn and disappearing in the spring; it is also more confined to the north than the Carrion Crow : in colouring it differs from the latter in the smoky-grey tint of the nape, back, rump, and lower parts of the body, except the black feathers covering the tibio-tarsal joints, and in its dark, horn-coloured claws.*

In Siberia the Hooded and Carrion Crows were discovered by Mr. Seebohm to interbreed. He says—" As you travel eastward from Tomsk, for about 120 miles, the Hooded Crow only is to be seen on the roadsides, and during the last 120 miles before reaching Krasnoyarsk the Carrion Crow alone is found. But in the intermediate hundred miles . . . . one fourth of the Crows are thorough-bred Hoodies, one fourth are pure Carrion Crows, and the remaining half are hybrids of every stage."—(Hist. Brit. Birds, vol. i., p. 547.)

---

* I take the above distinction from the 4th edition of Yarrell's 'History of British Birds.'

ROOK.
⸺ CORVUS FRUGILEGUS, *Linn.*
Pl. XV., figs. 7, 8.

Geogr. distr.—Europe generally; Asia as far east as India: common everywhere in Great Britain.
Food.—Seeds and grain, fruit, worms, slugs, insects in all stages, fish, young birds, mice, and carrion.
Nest.—A bulky structure of rough sticks, lined with mud, straws, grass, wool, fibrous roots, &c.
Position of nest.—In the upper branches of tall, growing trees; many nests being usually built close together, and forming what is known as a rookery.
Number of eggs.—4-5.
Time of nidification.—III-IV.

From the observation of some years I am led to the conclusion that the Rook seldom nests earlier than March, at any rate in the counties of Kent and Surrey.*

In severe winters it is not a rare occurrence for a Rook to stoop suddenly upon some unfortunate Sparrow, which it kills with a blow of its bill, and, putting one foot upon it, pulls it to pieces after the manner of a Hawk. I have on several occasions witnessed this predatory habit, but always on the coldest of winter days, when probably no other food could be obtained. At such times the Rook will come into one's garden and feed upon fat, pie-crust, or other table refuse, driving off the smaller birds; no sooner, however, does it take wing than these same pigmies mob it, and frequently succeed in so far annoying it that the choice morsel is dropped, and seized upon by the pursuers. The same thing may constantly be observed in the spring, when the Rooks are chased by Starlings, which rob them of almost every grub that they attempt to carry to their nests, so that it is a matter for wonder how the young Rooks get sufficient to sustain life.

The note of this bird is a long-drawn " *carr.*"

---

* It is said not to commence building in earnest until from the 10th to the 11th March; but this seems to me rather too precise.

JACKDAW.

CORVUS MONEDULA, *Linn.*

Pl. XVI., figs. 1-4.

Geogr. distr.—Europe generally, but rather local; western Asia and northern Africa : common and resident in Great Britain.

Food.—Insects, worms, Mollusca, reptiles, birds' eggs, carrion, seeds, peas, and fruits.

Nest.—Bulky and careless; the foundation of sticks and straws, upon which are heaped feathers, wool, flue, or any soft rubbish.

Position of nest.—In holes and clefts of cliffs, old church towers, ruins, chimneys, &c.; has been known to build in trees.

Number of eggs, 4-7 ; rarely more than 6.

Time of nidification.—V.

The Jackdaw is rather inclined to be gregarious, four or five nests being frequently found in the same church-tower; whilst in the holes in the cliffs at Dover many pairs breed every year, so that in the evening one hears their sharp cry—" *chark* "—incessantly. This sound has doubtless suggested the name " Jack," by which the bird, whether wild or domesticated, is universally known in our country.

As a pet the Jackdaw is extremely entertaining, but at the same time is always on the look-out for a piece of mischief; he is also an incorrigible thief, and if any glittering object comes under his notice he will watch for an opportunity to seize and hide it : in avoiding a missile or catching a dainty morsel his dexterity is marvellous, and the dodging or catching appears to be done without the least effort; it is never overdone.

A gentleman with whom I sometimes travel found a nest of a Jackdaw built in part of an old chimney-pot, and consisting wholly of flue, probably collected from the sweepings of the house : such as I have seen in niches in old church towers have usually consisted chiefly of straw and feathers upon a few rough sticks ; they could hardly be called nests, having more the appearance of a collection of rubbish loosely thrown together as a bed for the eggs.

MAGPIE.
PICA RUSTICA, *Scop.*
Pl. XVI., figs. 5, 6.

Geogr. distr.—Entire Palæarctic region ; common in Great Britain.
Food.—Insects in various stages, birds' eggs, young chickens and other small birds, reptiles, mice, carrion, grain and fruit.
Nest.—A large bulky construction of sticks (which are for the most part thorny), the true nest covered by a strong dome, and itself formed of fine roots and dried grass on a foundation of earth, which is plastered upon sticks forming the bottom of the external structure.
Position of nest.—Either high up in a tree or low down in a bush or hedge ; it is also said sometimes to build under eaves, or even on the ground.
Number of eggs.—6-8.
Time of nidification.—IV-V.

The Magpie is said to be very destructive to eggs of game ; it unquestionably is to eggs of the smaller birds, such as the Blackbird and Thrush. It also sometimes robs poultry-yards and preserves of young chicks, and thus brings swift destruction upon itself. Nevertheless it is still an abundant species in the county of Kent, and I have frequently seen several pairs at a time flying off at my approach in a small wood near Newington on the Chatham and Dover line. The Jay is equally common in the same wood, but I have not seen the nest of either species there, though I have known of nests of the Pie within two miles of that place.

When domesticated the Magpie exhibits the same nature as the Jackdaw, pilfering any shining object, destroying fancy work, tormenting any dog or cat small enough not to be feared, and generally proving itself what is known as a "delightful torment." The Magpie, the Jackdaw, or the Chough, when petted, are as amusing, and, at the same time, require as much watching, as a monkey; like that animal, they are ever on the look out for self-gratification, no matter at what cost to their fellow creatures.

COMMON JAY.
GARRULUS GLANDARIUS, L*inn*.

Pl. XVI., figs. 7, 8.

Geogr. distr.—From the central and northern parts of Scandinavia, throughout Europe to Algeria, and eastward as far as the Ural Mountains : common and resident in Great Britain.

Food.—Insects, worms, birds' eggs, young birds, grain, berries, haws, acorns, beech-nuts and peas.

Nest.—Externally rather ragged, being formed of sticks and twigs intertwined ; more compact and neatly cup-shaped in the centre, which is lined with rootlets and horse-hair, and sometimes grasses.

Position of nest.—Usually in a high bush or the fork of a sapling ; according to my experience, most frequently in rather close growing woods, without undergrowth, so that it is readily seen without search : Mr. Dresser, however, says that the nest is " well concealed," so that it is possible I may have only discovered such as were easy to find.

Number of eggs.—6-7.

Time of nidification.—III-VI ; April and May.

The Jay is a common bird in Kent, and, considering how many one sees during the breeding season, I have often wondered at the few nests which I have found. This bird is, however, generally admitted to be less common in England than formerly, though it is said to have increased of late years in Lincolnshire. In Scotland, however, it has been proved to have decreased rapidly, being rare in most counties south of the Grampians, and local in all ; in Ireland, also, it appears to be confined to the south, and to be very local and by no means numerous.

The Jay is a very timid bird, and usually flies noisily off, or conceals itself in the surrounding foliage at the approach of man. The first nest which I took was apparently unguarded, though it contained six eggs ; but the moment I removed them the scolding I received made me aware of the vicinity of the parent bird. This species has a remarkable power of imitation, and has been known to mimic the crow of a Cock, the cackle of a Hen, the hoot of an Owl, the bark of a Dog, or the neigh of a Horse ; its natural note is harsh and discordant, but it can utter more melodious sounds when so inclined.

Family CUCULIDÆ.
CUCKOO.
Cuculus canorus, Linn.
Pl. XVI., figs. 9-11.

Geogr. distr.—Nearly the whole of Europe and Northern Asia in summer; towards winter, however, it migrates to S. Africa and India.
Food.—Insects, Mollusca, berries.
Nest.—None is constructed, the eggs being placed in the habitations of other birds, which act as foster parents to the young Cuckoo.
Number of eggs.—As only one is placed in each nest, this is difficult to ascertain: but the number probably varies from 4.8.*
Time of nidification.—V-VI ; May.

The Cuckoo usually arrives in this country about the middle of April, and leaves again in August, the young birds following about a month later : its favourite nurseries appear to be the nests of the Hedge Sparrow, Reed Warbler, Sedge Warbler, Pied Wagtail, Tree and Meadow Pipits, and Skylark. I have myself taken it in all of these excepting the Sedge Warbler and Meadow Pipit, as also in that of the Yellowhammer, and I have seen the young bird nearly ready to fly in that of the Song Thrush (Zool. 1877, p. 300). In addition to the above it has been found (See Dresser's 'Birds of Europe') in nests of the Red-backed Shrike, Spotted Flycatcher, Blackbird, Ring Ouzel, Wheatear, Stonechat, Redstart, Robin, Grasshopper Warbler, Dartford Warbler, Whitethroat, Lesser Whitethroat, Garden Warbler, Blackcap, Wood Wren, Chiff-Chaff, Willow Wren, Common Wren, Grey Wagtail, Yellow Wagtail, Rock Pipit, Woodlark, Common Bunting, Reed Bunting, Cirl Bunting, Chaffinch, Linnet, and Swallow ; in several of which nests it could only be deposited by being taken up in the bill of the bird and dropped thence into its place, as has been recorded by several recent observers to have been seen done.

In 1884 I came across a singular instance of deficient instinct on the part of a Cuckoo, which deposited an egg in an unfinished Linnet's nest entirely destitute of lining; I watched the nest for several days, and, finding that it was deserted by the Linnets, I took it as a curiosity. The same year (June 19th) Mr. Hickling (of Sidcup, in Kent) took a nest of the Pied Wagtail containing two Cuckoos' eggs from ivy on his garden wall, and sent it to me ; both eggs were deposited on the same morning, and therefore by two Cuckoos.

* A friend of mine took five in a small swampy grove, in one evening; all were in nests of the Sedge Warbler; they probably represented the complete laying of one female.

Family HIRUNDINIDÆ.

SWALLOW.

Hirundo rustica, *Linn.*

Pl. XVI., figs. 12, 13; and Pl. XXXVII., fig. 6.

Geogr. distr.—Throughout Europe, Asia, and Africa; a common summer visitant in Great Britain, arriving in April and departing in October or November.

Food.—Insects.

Nest.—Semicircular or demisemicircular in shape, according to the position in which it is placed; open at the top; the walls are thick, and formed of mud pellets mixed with straw or hay; the lining is of fine grass-stems, generally almost hidden by a quantity of feathers, though these are sometimes absent.

Position of nest.—Inside chimneys, usually three or four feet from the top, though sometimes not so low down; in corners of stone vestibules, in barns or sheds, built upon a beam, or in a corner at the junction of the rafters; also down the shaft of an old well or deserted mine.

Number of eggs.—4-5; rarely 6.

Time of nidification.—V-VIII; May, and rarely later than July.

The Swallow is double-brooded, the first sitting being generally complete about the second or third week of May. The same nest is often made use of for two successive years, when sufficiently protected from the weather for its preservation. I have a nest in my possession taken from the corner of the vestibule of a large house, along with the eggs laid in it in two successive years; the first clutch was somewhat abnormal in pattern, and I have figured one of the eggs on my supplementary plate (Pl. XXXVII., fig. 6).

When approaching the nest the parent birds always give notice of their advent by a sharp and somewhat plaintive double note, doubtless to prepare their young to be ready for meals, though the old birds evidently do not know why they do it, since they behave in the same way when the nest is empty, or only contains eggs. This is, however, unfortunate for the young oölogist, who is often told that the birds are crying for the loss of the eggs which he has deprived them of, and is forbidden to continue his favourite pursuit; yet those who reprove him will unblushingly rob the poor domestic fowl every time she lays.

In the 'Field' for 1871, p. 281, Mr. Harting has recorded the shooting of a white variety of this bird, but as the iris was of the usual brown colour, the crown exhibited a few dusky feathers, and the throat was pale rufous, it could not be called a pure albino; the bill and interior of the mouth were bright yellow.

MARTIN.

CHELIDON URBICA, *Linn.*

Pl. XVII., figs. 8, 9.

Geogr. distr.—Europe generally in summer; Africa in winter; eastward as far as Persia: generally throughout England, Scotland, and Ireland.

Food.—Insects.

Nest.—A semicircular structure formed of mud pellets mixed with straw, and with a small opening in front at the top; lined with grass-straws, wool, feathers, and hair.

Position of nest.—Fastened under porches, archways, or eaves, over windows; or under window-sills or ledges projecting from the face of a rock.

Number of eggs.—4-5.

Time of nidification.—V-VIII; June.

I have always found the nest of this species more brittle than that of the Swallow; those which I have examined contained less admixture of straw with the mud, and were thinner as well as deeper than those of *H. rustica,* so that, whereas the nest of the latter could be detached without injury by letting down a metal dipper full of hay below it, and simply jerking it upwards, the nest of the Martin, though carefully held in front whilst a knife was passed round the edges, invariably got broken in removal.

The Martin is a sociable bird, and it is not uncommon (at any rate in Kentish villages) to see six or eight nests in a row under the eaves of a house, two or more of them some-times close together. The eggs are said to be laid towards the end of May, but I am convinced that this is rarely the case, at any rate in the South of England, where I have usually been unable to obtain eggs before the second week in June, whereas eggs of the Swallow can be obtained in the first week of May. The Martin arrives in this country in April, about a week later than the Swallow: commences to build in May (generally about the middle of the month), and the completed nest sometimes remains about eight days before the first egg is laid, but I cannot say that this is always the case, as I have only twice had an opportunity of ascertaining the fact.

The song of the Martin is inferior to that of the Swallow, and consists of a low twittering mingled with a few melodious notes.

SAND MARTIN.

COTILE RIPARIA, *Linn.*

Pl. XVII., figs. 10, 11.

Geogr. distr.—Europe, Asia, N. Africa, and America as far south-ward as Brazil: common and generally distributed in Great Britain.

Food.—Insects.

Nest.—Loosely constructed of hay, rootlets, and a little vegetable fibre; lined with feathers, which are almost invariably all white.

Position of nest.—In an angle or shelf at the end of a long hole bored by the bird itself in the walls of sand or gravel-pits, or the sand-bank of a railway cutting or sunken road.

Number of eggs.—4-5; usually 5.

Time of nidification.—V-VI; beginning of June.

This species arrives in Great Britain late in April or early in May, and soon afterwards commences to excavate its holes, or deepen those already made : some of these holes extend so far into the sand that it is impossible to reach them without the help of a long fork; I have used a telescopic toasting-fork for this purpose, but it is apt to break the eggs; a small, long-handled metal bowl lined with wool would be far better. Sometimes, however, the nest is quite out of reach. I have buried my arm to the shoulder, with a wire over two feet in length held between my fingers, and have not been able to touch the end of the tunnel.

The holes made by the Sand Martin are frequently taken possession of by other birds, but especially by House Sparrows and Starlings, so that they get filled with rubbish; the Blue Tit and Kingfisher also are not averse to utilizing them; the Martins are therefore compelled to unnecessarily multiply their tunnels, thus not unfrequently causing a land-slip during the succeeding winter.

As a rule this species is gregarious in its habits, but instances have been recorded of one pair nesting by itself.

Mr. Seebohm (Hist. Brit. Birds, ii., p. 185) mentions that he has seen Sand Martins flying in and out of tunnels which they had excavated in enormous heaps of half-rotten sawdust.

Family CYPSELIDÆ.
COMMON SWIFT.
Cypselus apus, *Linn.*
Pl. XVII., figs. 12, 13.

Geogr. distr.—Europe generally in the summer; Asia as far east as Dauria; Africa in the winter: it arrives late in Great Britain, and leaves early.

Food.—Insects.

Nest.—Formed roughly of straws, grasses, feathers, moss, wool, and cotton, felted and glued together with a viscid secretion from the mouth.

Position of nest.—In crannies in cliffs, ruins, old church-towers, thatches, hollow branches of decayed trees, &c.

Number of eggs.—2-4.

Time of nidification.—V-VI.

The Swift arrives in the Southern counties towards the end of April, and in the northern counties early in May; it usually leaves again at the end of August or early in September; though less common than the Martins and Swallow, I remember to have formerly seen it in some abundance in the village of Herne, in Kent, and specimens caught at Herne Bay were sometimes brought into the hotel.

The nest is scanty and rather flat, carelessly built, but the materials covered with a viscid saliva which unites them firmly together. It is frequently placed under the roofing of a house in a similar position to that usually affected by the House Sparrow. The number of eggs laid is usually two, rarely as many as four, of a pure chalky-white colour.

It was formerly supposed to be unable to rise from the ground, and it certainly prefers to fly from a higher level; I have, however, seen a Swift rise from the high road at the approach of some village boys who made sure of capturing it, and apparently without any effort. Taken into a house and placed upon a carpet, the Swift tumbles about, owing, probably, to the fact that the woollen fibres get entangled in its claws as it attempts to rise. I remember one bird which, after tumbling about, fell down apparently dead, and when thrown out of the window dropped like a stone till within a foot or two of the ground, when in a moment its wings opened and it sailed swiftly away.

FAMILY **CAPRIMULGIDÆ.**

COMMON NIGHTJAR.

CAPRIMULGUS EUROPÆUS, *Linn.*

Pl. XVI., figs. 14, 15.

Geogr. distr.—Europe generally in the summer; as far eastward as
Persia and Turkestan; migrating southward to Africa at the approach
of winter: in suitable localities throughout Great Britain.
Food.—Insects.
Nest.—A mere depression in the ground.
Position of nest.—Either on open heaths or amongst trees.
Number of eggs.—2.
Time of nidification.—VI.

The Nightjar arrives in Great Britain early in May, and
leaves again in September. "It is," says Seebohm, "a
late breeder, and its eggs are not usually deposited before
the beginning of June. In some seasons they may be
found as early as the end of May, but this is exceptional.
It makes no nest, and deposits its eggs upon the ground,
sometimes at the foot of a tree, in rare instances on a
fallen trunk covered with moss and lichen, often in a slight
depression on the far-stretching heathy wastes, but most
commonly on a small, naked, flat patch of ground amongst
the bracken and the brambles. Here the female deposits
her two eggs; and as incubation advances a little hollow is
often worn into the earth by the incessant sitting of the
bird, but no preparation is ever made. Only one brood
appears to be reared in the year; but if the first clutch of
eggs is taken or destroyed others are usually laid, and this
accounts for the late eggs of the species that are sometimes
found in July, and even in August."—(Hist. Brit. Birds, ii.,
p. 313.)

J. H. Gurney says:—"It has been doubted whether the
Nightjar rears two broods in a season; that it generally
does so in Norfolk I feel sure, the contrary opinion having
perhaps arisen from the circumstance of its being so late a
migrant. That the eggs at Causton" (taken 4th August)
"were a second laying by birds which had had young
previously, I think, as I saw four young ones at the same
place able to fly on the 19th of the month previous (July).
These would have been at least twenty-one days old (most
likely older), and two of them were probably the first brood
of the pair whose eggs Mr. Norgate and I found on
August 4th."—(Zool., 1883, pp. 429-30.)

FAMILY PICIDÆ.
SUB-FAMILY *PICINÆ*.
GREEN WOODPECKER.
GEOINUS VIRIDIS, *Linn.*
Pl. XVII., fig. 1.

Geogr. distr.—Throughout Europe, but not extending into Siberia: more common in the southern than the northern counties of England.

Food.—Insects in all stages; probably also nuts and acorns.

Nest.—A mere hole cut through the sound outer wood of a tree into the decayed portion; the hollow forming the nest is a few inches below the entrance, or some distance into the tree, and often far from the ground, and the eggs are deposited upon the rotten wood.

Position of nest.—In decaying oaks in forests, or ashes and elms in hedge-rows; also in poplars, horse-chestnuts, sycamores, silver-firs, and beeches.*

Number of eggs.—5-7; rarely 8.

Time of nidification. - IV-VI; early in May.

I have had some difficulty in ascertaining the time of nidification of this species. Eggs in my collection were taken in the New Forest in June. It is strange that books intended to guide the collector of eggs should frequently omit altogether to indicate when the various species should be sought for. The three principal needs of a collector are represented by the words *where, when,* and *how,* and in giving answers to all these questions Mr. Seebohm's 'History of British Birds' stands first. From his book I cull the following :—

" The usual note of this bird, which is uttered most frequently in spring and early summer, is a loud and clear *kyu, kyu, kyu,* so rapidly repeated as to bear a great resemblance to a hearty laugh. Its loud cries have been said to prognosticate rain ; and in many districts the Green Woodpecker is known as the ' Rain-bird,' whilst its singular note, has gained for it the name of ' Yaffle' in some counties. When alarmed the note of this bird is somewhat modulated, and resembles, perhaps, less a laugh than a scream " (p. 366.)

This Woodpecker not only feeds on timber-haunting insects, which it picks from the bark of the trees, commencing near the bottom of a trunk or branch, and working upwards in an oblique direction, on larvæ, and spiders, but in summer it destroys great numbers of ants, visiting one nest after another, boring into the ground and securing the ants as they fall in. It is also said to prey upon bees.

---

* According to Newton, the oak and beech are rarely chosen.

GREAT SPOTTED WOODPECKER.

Picus major, *Linn.*

Pl. XVII., fig. 2.

Geogr. distr.—Throughout Europe and Siberia to Japan, and probably the entire Palæarctic region: in the southern and midland counties of England and in Scotland. It is still scarce, though apparently less so than formerly.

Food.—Insects in various stages, berries, nuts, acorns, and fruits.

Nest.—A mere circular hole bored twelve to eighteen inches into a tree.

Position of nest.—In the branch or trunk of a tree the heart of which is decayed.

Number of eggs.—5-8; usually 5.

Time of nidification.—IV-V; middle of May.

This species, though less rare in some localities than others, is by no means so common as the Green Wood-pecker; nevertheless it is supposed to breed in every county in England and Wales. According to Newton (Yarrell's Hist. Brit. Birds, 4th ed., p. 471) "it seems seldom, if ever, to inhabit precisely the same spots as the Green Woodpecker; yet its haunts are very varied in character—large oak woods, hedgerows where ashes form the prevalent timber, holts or small plantations of poplars and alders, and the lines of pollard-willows that skirt so many rivers. In many of its habits—its solitary and mis-trustful disposition, its mode of flight, and of climbing—it closely resembles its larger relative; but it usually affects trees of smaller growth, and more frequently alights and seeks its food on the upper branches than on the trunk, and indeed, would seem sometimes to sit crossways on a bough after the usual fashion of birds. It is rarely seen on the ground."

Seebohm says that the hole for the nest "is often made where a branch has been blown away and the rain has rotted a small hole into the trunk. This is often enlarged if it be not already big enough for the purpose. The hole varies in extent, sometimes being as much as eighteen inches deep, but frequently only a foot, and, more rarely, the eggs are within reach of the hand. The passage is wonderfully round and smooth, and the end is enlarged a little into a sort of chamber, and here the eggs are deposited. The bird makes no nest; the eggs lie upon the powdered wood at the bottom of the hole."—(Hist. Brit. Birds, vol. ii., p. 357.)

LESSER SPOTTED WOODPECKER.

Picus minor, *Linn.*

Pl. XVII., fig. 3.

Geogr. distr.—Europe generally; occurs sparingly in England, being less rare in the southern counties, " where oaks, aspens, ash, and other non-evergreen trees, are scattered through the conifer growth ;* it has been known to breed in Wiltshire.

Food.—Insects and their larvæ.

Nest.—A round hole 1¼ inches in diameter, and about a foot deep, terminating in a circular cavity; eggs deposited upon rotten wood.

Position of nest.—In the trunk of a decayed tree at a considerable height above the ground.

Number of eggs.—4-6.

Time of nidification.—IV·V.

This is a very shy bird; but when searching for insects it has been known to allow itself to be closely approached. It is particularly active in its movements, which are not unlike those of a Creeper.

According to Newton, "the Western Midlands, the counties of Gloucester, Hereford, Salop, Worcester, and Warwick, appear to be its chief resort. Cornwall perhaps excepted, it breeds in every English county as far as York, but there becomes very rare, and is only a casual visitor in Lancashire or to the northward."—(Yarrell's Hist. Brit. Birds, 4th ed., vol. ii., p. 480.)

" Its hole," says Seebohm, " is made in many kinds of trees, and at different heights from the ground. Sometimes it chooses a dead stump, or the stem of an apple or a pear tree, more frequently high up in the branches of a poplar, a beech, or an elm. Sometimes it bores into a pollard willow by the stream, or selects a pine or birch tree for its purpose. The hole is bored by the industrious little miners for a distance of a foot or more (sometimes only eight or nine inches). The hole is round, gradually enlarges as it proceeds downwards, and at the extremity widens out into a small hollow, where the eggs are laid. The diameter of the passage to this chamber varies from about an inch and a half to two inches.—(Hist. Brit. Birds, vol. ii., p. 361.)

As with the Greater Spotted Woodpecker, the eggs are laid upon the chips and powdered wood at the extremity of the hole.

---

* Dresser's ' Birds of Europe.'

SUB-FAMILY *JYNGINÆ.*
WRYNECK.
JYNX TORQUILLA, *Linn.*
Pl. XVII., fig. 4.

Geogr. distr.—Europe generally as far as N. Scandinavia during
the summer; eastward to Japan; migrates to Africa for the winter:
common in England, rare in Scotland, unknown in Ireland.

Food.—Insects.

Nest.—A mere hole in a tree, the eggs being laid upon the decayed
wood; has been known to lay in Sand Martins' burrows.

Position of nest.—In hole in apple, pear, or soft-wooded non-evergreen tree in orchards, gardens, or well-wooded localities; seldom more
than five or six feet from the ground.

Number of eggs.—5-12; usually 9-10.

Time of nidification.—V.-VI.

Two or three years ago, on visiting an orchard a few
days previous to my return to town, I observed a Wryneck
flying about a decayed old apple tree in which were several
holes; one of these holes had been occupied the previous
year by a Robin, and a little withered grass still remained
in it; I called a boy and promised him a shilling if he
would send me the whole of the eggs as soon as the hen
ceased to lay. A little more than a week after this I
received five eggs, which were deposited upon the remains
of the Robin's nest; no more eggs were laid after the fifth.

As is the case with the Blue Tit, this bird hisses angrily
when disturbed upon the nest, a habit which (in Kent) has
procured it the name of the Snake Bird. Its cry is a sharp
whistle, supposed to represent the word "*jynx,*" whence its
generic name.*

The Wryneck arrives in this country about the end of
March, or the first or second week of April, and is most
commonly met with in old orchards or gardens bounded by
decayed willows or elms. It is quite fearless, but when
taken in the hand shows much anger, hissing, erecting its
head-feathers, and pecking fiercely, at the same time
grasping one's hand with its claws, but making no effort to
escape.

Ants form the principal food of this bird, but many
other insects are devoured by it.

---

* It has also been described as " *qui, qui, qui, qui, qui, qui, qui,*"
in rapid succession : and Seebohm says it bears some resemblance to
the word " *vite* " uttered several times in succession.

Family UPUPIDÆ.

HOOPOE.

Upupa epops, Linn.

Pl. XVII., fig. 5.

Geogr. distr.—Common in S. Europe, rarer in the North; occurs also in China, and is a winter visitant to India and Africa: in Great Britain it is by no means common at the present time, though it would appear that formerly it was not an uncommon species.

Food.—Insects and worms.

Nest.—Frequently none, but sometimes a few bents, feathers, twigs, and a little cow-dung.

Position of nest.—In decayed and hollow trees, the eggs being usually deposited upon the rotten wood; rarely upon the ground, under a large stone, or in holes on the sunny side of walls or rocks.

Number of eggs.—5-8; rarely more than 7.

Time of nidification.—V.-VI.

The Hoopoe has been described as flying somewhat in the manner of the Lapwing. Seebohm says that when leaving the nest the flight has a butterfly appearance, and anyone who has seen the bird can understand that on the wing, especially when near, it would much resemble butterflies of the tropical genus *Neptis*, though on a much larger scale.

The nest, which is always in a hole, is described as being extremely offensive, the bird not only being of an uncleanly habit itself, but evidently a lover of abominations. Newton says that "the furnishing of its nursery is nearly always completed by introducing some of the foulest material that can be conceived."

Several instances have been recorded of the nesting of this bird in England, but unfortunately its conspicuous appearance, combined with its great timidity, militate against the success of most of its attempts to do so. It can hardly be considered a rare bird in this country, for specimens are shot every year, and it has been observed in most of the counties of Great Britain.

The colouring of the eggs, when fresh, is said to vary considerably, some being pale greenish blue, somewhat resembling the usual colour of the Starling's egg (though, by the way, the colour of this egg varies from bluish white to a colour only a shade lighter than that of the egg of the Song Thrush), whilst others are lavender-grey, or even stone-colour.

Family MEROPIDÆ.

COMMON BEE-EATER.

Merops apiaster, Linn.

Pl. XVII., fig. 6.

Geogr. distr.—Southern Europe generally in summer; Western Asia and throughout Africa to the Cape of Good Hope : rare in Great Britain.
Food.—Insects and berries.
Nest.—A mere chamber hollowed out of the earth, and about a foot in diameter, at the end of a horizontal hole (or tunnel) three to four feet in length, and gradually enlarged from the entrance, the eggs being deposited upon the soil.
Position of nest.—In a bank, sand-hill, or cliff, usually overhanging a stream, though sometimes at a distance from water.
Number of eggs.—5-8 ; usually 6.
Time of nidification.—VI ; early in June.

This species arrives in April or May, and leaves again in July, August, or September, though specimens are noted as having been shot in Great Britain as late as October. It appears to be extremely rare in this country, and there-fore, though a summer visitor, it is hardly surprising that only one instance of its supposed nesting in our country is recorded ; moreover, of those birds which have been obtained, as Newton has pointed out (Yarrell's Hist. Brit. Birds, 4th ed., p. 438), " most of the dated captures have occurred between the end of April and beginning of May, so that a majority of the examples which have visited us have doubtless been seeking a breeding quarter."

Mr. Seebohm, in his ' History of British Birds,' vol. ii., pp. 323-4, thus describes the taking of the nest of this species :—" When at Kustendji last year, in company with Mr. Young, I dug out, on the 15th of June, half-a-dozen of their nests from the old Russian trenches formed during the last war between that power and Turkey; they were from two to four feet from the surface, and penetrated about four feet into the ground, nearly straight and nearly horizontal. Two of the nests we dug out contained eggs, one four and the other six, all fresh. The old nests con-tained nothing but fine earth and decomposed castings. The eggs in the four clutch were on the bare earth ; those in the six clutch were surrounded with beetle-cases and wings of dragonflies. The latter did not appear to have been swallowed ; so that it seems probable that the male feeds the female on the nest. The birds sat very close, and did not leave the hole until it was half dug out."

Family **ALCEDINIDÆ.**

COMMON KINGFISHER.

ALCEDO ISPIDA, *Linn.*

Pl. XVII., fig. 7.

Geogr. distr.—Temperate Palæarctic region generally: in Great Britain it occurs everywhere (but not in any numbers), and is resident.
Food.—Insects, Crustacea, Mollusca, fish, frog-spawn.
Nest.—A hole in a bank, round externally, about two feet deep, larger at the extremity: the eggs being deposited upon a collection of fish-bones and castings.
Position of nest.—Usually near to, or over-hanging water, though sometimes at some distance from water.
Number of eggs.—5-7.
Time of nidification.—IV-VII,* but only one-brooded.

Although I have on several occasions known of the existence of the nest of the Kingfisher in private grounds, I have never been able to take it, owing to the beauty of the bird and its comparative rarity, which endear it to those in whose property it takes up its abode, and who preserve it with even more jealousy than they would game. On the wing it certainly is a perfect gem, but its nesting habits are decidedly offensive, though, as is usual with birds which nest in holes, the eggs are beautifully white, and, as with many species which make no regular nest, are extremely hard and smooth, so that, until the first puncture is made, a steel drill is apt to slip upon their surface unless handled with care.

The Kingfisher is rarely seen in company, but towards winter a pair will occasionally be seen together; in 1884 I had a pair pointed out to me flying over the tanks at Battersea, and the same day I saw a third at Dulwich. The flight is swift, but regular, so as to exhibit the full beauty of the bird as it passes.

In 1885, my friend Mr. Salter obtained the nest of a Kingfisher from a Sand Martin's hole ; but as it consisted, as usual, of nothing but a collection of fish-bones, he did not keep it for me. As a rule the bird prefers to build close to water, over which it delights to fly. Its food consists principally of fish.

---

* Hewitson expresses the opinion (Zool. 1867, p. 707) that it probably breeds as early as the beginning of March ; it has been known to lay in February.

Family **COLUMBIDÆ.**

RING DOVE.

COLUMBA PALUMBUS, Linn.

Pl. XVII., fig. 14.

Geogr. distr.—Europe generally, excepting in the extreme north, not ranging far into Asia: common, widely distributed, and resident, in the British Isles.

Food.—Clover, young turnip and cabbage leaves, peas, berries, acorns, mast and grain.

Nest.—A platform of twigs, rather open in construction, or an old Crow, Jay, or Squirrel's nest.

Position of nest.—High up, as a rule, in trees near the edge of a wood, or in the middle of a forest or garden.

Number of eggs.—Usually 2; occasionally 3.

Time of nidification.—IV-VI.

It is said that during severe and protracted frosts the Ring Dove will feed voraciously upon the tubers of the wood anemone.

A nest which I took at Kemsley, in Kent, in 1885, is constructed entirely of roots, some trees having recently been cut down and the roots left lying about. The nest was placed on a kind of platform in the branches of a weather-beaten old hawthorn, so gnarled and twisted that the boughs formed a series of steps up to the nest. It contained only one egg, slightly incubated. On the other hand, Mr. Janson discovered a nest high up in the ivy clinging to a tall straight tree much frequented by this bird, in which nest were three helpless young ones.*

The Ring Dove differs from our other species by its superior size and white collar.

In approaching a nest the position of which one is anxious to discover by the sudden flight of the bird from it, it is of some use to arrest its attention and partially allay its fear of intrusion by imitating its note. To some, however, this is not easy of accomplishment, as the sound is produced by a combination of the sound " *oo* " with a gargle, but probably many could, with practice, become perfect in it.

---

* I consider this fact especially worthy of note, as it is constantly asserted that the Ring Dove never lays more than two eggs; I was present at the finding of the nest, and the three young ones were lifted up and shown to me.

STOCK DOVE.
COLUMBA ÆNAS, *Linn.*
Pl. XVII., fig. 15.

Geogr. distr.—Southern Europe, N. W. Africa, eastward as far as Persia: in Great Britain chiefly confined to the midland and eastern counties of England.

Food.—Young leaves of vegetables, peas, berries, acorns, mast, grain and other seeds.

Nest.—Constructed of thin twigs or stems of plants, sometimes intertwined with rootlets, dried leaves, and a little moss.

Position of nest.—In holes of old trees, in ivy clinging to trunks of cedars and firs, in fir trees and pollard hornbeams, and sometimes on the sand in rabbit-burrows about a yard from the entrance, or in crevices in rocks.

Number of eggs.—2.

Time of nidification.—IV-VII; sometimes as late as October.

This species, the almost uniform plum-colour of whose plumage distinguishes it from our other doves, is believed to be increasing in numbers in Great Britain : though local, it is to be found throughout England and Wales; in Scotland and Ireland it is very rare.

"The Stock Dove," says Seebohm, "can scarcely be regarded as a forest bird, though it is especially partial to well-timbered parks. It spends nearly all day in the open country, but frequents the skirts of the forests in order to find a breeding-place in the hollows of the old trees. It frequents the flat, open country of the lowlands, where the pollard willows provide it with a suitable nesting-site, and makes its home both on the stupendous sea-girt cliffs and the limestone crags or quarries of the moors. The portion of Sherwood Forest known as Birklands is a paradise for the Stock Dove, abounding as it does in hollow old oaks. It frequents this district throughout the year, and may be repeatedly seen flying to and from the woodlands to the neighbouring fields."

"In places where there are no hollow trees the Stock Dove often rears its young in the old nest of a Magpie or Crow; or in the dense ivy growing over trees or buildings." —(Hist. Brit. Birds, vol. ii., p. 402).

In Hewitson's nesting days this species appears to have been tolerably common in Epping Forest, and he mentions having taken its eggs from pollard hornbeams.

ROCK DOVE.
COLUMBA LIVIA, *Bonn.*
Pl. XVII., fig. 16.

Geogr. distr.—Northern Scandinavia to N. Africa, eastwards to China and Japan : fairly common in suitable localities in Great Britain.

Food.—Young leaves of vegetables, peas, berries, grain, mast, Mollusca

Nest.—Carelessly constructed of withered grass, sprigs of heather, or seaweed.

Position of nest.—In caverns, crevices, and holes in cliffs and rocks upon the coast.

Number of eggs.—2.

Time of nidification.—III-IX ; rarely as late as October.

This is the original of the domesticated pigeon; it may always be recognised in a wild state by its white rump. Hewitson says that Rock Doves abound in the Shetland Islands, " breeding in the numerous spacious caverns, into which the sea constantly rushes ; they have every appearance of being tame, and are so easily approached within gun-shot, that, until assured of the contrary, we took them for the inmates of some neighbouring dovecote. They approach quite close to the huts of the fishermen, to feed over the small cultivated patches of corn-land."—(Ill. Eggs Brit. Birds, vol. ii., p. 228.)

Seebohm says, " The true home of the Rock Dove is on the rocky coasts—bold headlands and beetling cliffs which are tunnelled by ocean caves. Cliffs where there are few caves are not so much frequented by the Rock Dove ; thus at Hamborough, although it is far from uncommon, it is not nearly so numerous as on the wild, rugged, western shores of Scotland, where so many caves are found." The note of this bird does not differ perceptibly from that of the Ring Dove, and is soft and full—*coo, coo, roo-coo.*"*
—(Hist. Brit. Birds, vol. ii., pp. 406, 407).

---

* This is quite as correct as words can render it; only, as the letter *r* is rolled in *F*rench, so the letter *o* is rolled in the language of Doves.

TURTLE DOVE.
TURTUR COMMUNIS, *Selby*.

Pl. XVII., fig. 17.

Geogr. distr.—Temperate Europe generally, migrating to Africa for the winter; also found in Western Asia; not at all common in the southern counties of Great Britain, somewhat rarer in the north, local and uncommon in Ireland.

Food.—Seed, peas, mast, nuts, and young shoots.

Nest.—A platform of twigs, bound together in the centre with roots; it varies considerably in size; one taken by myself, in 1883, measures as much as 12 inches by 16½ to the extremities of the twigs, whilst another only measures 7 inches by 9; the latter is somewhat smaller than the average of those which I have found.

Position of nest.—In woods or groves, frequently not far from water, and where the undergrowth is somewhat dense, but often in woods far distant from any visible stream (though not far from human habitations), and where the undergrowth consists merely of a little scattered bracken: the nest is generally placed in a fork, and is rarely so far from the ground as to necessitate climbing in order to remove the eggs.

Number of eggs.—Almost invariably 2; rarely 1.

Time of nidification.—V-VI.

I have generally found the nest of this bird, which is not uncommon in the Kentish woods, from the middle of May to the second week of June, but as a rule the eggs are more or less hard-set by the end of May, so far as my experience goes. I have found the bird by no means easily frightened, sitting apparently without concern upon its eggs until I had almost touched it, and then flying off without flurry or excessive noise. Indeed I have noticed that those birds which are ready at the first alarm to desert the nest invariably make the most fuss about it, just as bad mothers and wives always lament most loudly when they lose children or husband.

In a nest, containing one egg only, which I recently found, the egg was somewhat malformed; there was not space in the nest for two.

FAMILY **PHASIANIDÆ.**

PHEASANT.

PHASIANUS COLCHICUS, *Linn.*

Pl. XVIII., fig. 1.

Geogr. distr.—Temperate Europe generally; acclimatized and preserved in Great Britain.

Food.—Berries, seeds, especially cereals, oak-spangles, acorns, insects, Mollusca, worms.

Nest.—A mere depression scratched in the ground, and lined with roots, leaves, and dried grasses; rarely the deserted nest of an Owl is taken possession of.

Position of nest.—In long grasses, fern, or undergrowth, in woods, groves, or plantations, under hedge-rows or in open fields: it has been found at a height from the ground on the top of a stack or in an old Squirrel's drey.

Number of eggs.—8-14.

Time of nidification.—IV-VI ; occasionally later.

The pheasant is fond of the larvæ of *Tipula,* known to gardeners by the popular name of " niggers," and, according to Mr. Seebohm, " it consumes enormous quantities of wireworms. It is therefore a useful bird to the agriculturist. In Great Britain, to which it is not an indigenous species, though the date of its introduction is not known, it has been proved by Prof. Boyd Dawkins that it had become naturalized in England before the Norman invasion, and therefore, as Mr. Howard Saunders says, " it has undoubtedly maintained itself in this country in a wild state for a period sufficient to entitle it to be considered a British bird."—(Yarrell's Hist. Brit. Birds, 4th ed., vol. iii., p. 94.)

The introduction of the Chinese Ring-necked species or variety, *P. torquatus,* in the present century, has so far modified the British Pheasant that it is almost impossible, at the present day, to shoot a specimen which has not been more or less influenced by hybridization between the two forms. As with other gallinaceous birds, the Pheasant will breed with other species, and I have seen hybrids between it and the Golden and Silver Pheasants and the Common Fowl. It has also been known to breed with the Black Grouse and Guinea Fowl. Such hybrids, though they do not appear to breed when kept together, are productive when paired with either of the parent stocks.

The Pheasant with us is polygamous, but Mr. Seebohm believes that this is due to the semi-domesticated state in which it is found in England.

PARTRIDGE.

PERDIX CINEREA, *Lath.*

Pl. XVIII., fig. 2.

Geogr. distr.—Temperate Europe generally; rarer in the south (not occurring in N. Africa); does not extend far into Asia: generally in Great Britain.

Food.—Seeds, grain, berries, young leaves, Mollusca, worms, and insects.

Nest.—Usually a mere hollow in the ground, lined with a few dry straws or grasses; but occasionally a tolerably compact nest is made of these materials, apparently capable of being lifted out entire. I have twice seen such nests, whereas on another occasion I found the eggs in a bare hole into which water had oozed, so that they were partly imbedded in the mud.

Position of nest.—Under stunted bushes, in open places in woods or plantations, in open fields, under hedges, in grass-grown ditches, or hollow places in crumbling cliffs on the coast.

Number of eggs.—Usually 12-16; I have, however, known them to vary from 5-26, the latter number being deposited in three layers in a compact-looking nest on a small waste piece of ground at the junction of two country roads; they were covered with a little dried grass, which effectually concealed them, the nest being placed under a straggling bramble.

Time of nidification.—V.

It has been stated that the Partridge occasionally lays in the nest of the Pheasant. In two instances as many as thirty-three eggs have been found in one nest, but it is supposed that such large clutches are produced by more than one female; the fact that (so far as I have seen) large clutches are generally placed in a nest far more compact than usual, seems to point to the agency of two females in its construction.

Though readily tamed, this bird has rarely been known to breed in confinement. It is a strictly monogamous species, choosing its mate probably about the end of April, but, as Mr. Seebohm says, "the poor bird has so many enemies, that it seldom happens that a pair enjoy each other's society for many seasons in succession." When pursued by a dog it has been known (like the Pheasant) to take to the water and calmly puddle away with apparently as much ease as a Moorhen (see Zool., 1878, p. 349).

RED-LEGGED PARTRIDGE.

CACCABIS RUFA, *Linn.*

Pl. XIX., fig. 3.

Geogr. distr.—Madeira, the Azores, Western and Southern Europe; tolerably common in Great Britain, occuring as far north as Scotland.

Food.—Clover, leaves, berries, grain, seeds, insects, Mollusca, and worms.

Nest.—More perfect than is usually the case with the Common Partridge, composed of dried grasses mixed with feathers.

Position of nest.—Upon the ground in woods, and in heathy districts.

Number of eggs.—10-18.

Time of nidification.—V.

Although I have not personally found the nest of this species in England, I have on several occasions heard of its occurrence in the Kentish woods; but, as the keepers (I believe invariably) smash all the eggs which they come across, under the impression that, if allowed to breed freely, this bird would supplant the common species, it is not surprising that the Red-legged Partridge has become rare in this county; its occurrence in Norfolk and Suffolk is far more frequent.

Two or three years ago I received an egg of the Red-Legged Partridge, along with eggs of Terns collected for me in the vicinity of Romney Marsh.

In habits this species differs considerably from the Common Partridge; its flight is less noisy and more rapid; it has a disinclination to rise, and runs before a dog; all of which points are objectionable to the sportsman, whose pleasure seems to increase in proportion to the number of birds which he can either kill or maim, rather than to the skill required to obtain a fair shot; thus the crafty bird is despised and the silly victim esteemed. Unlike the common species, the males of *C. rufa* are said to desert the hens when they are sitting and rearing their young, and assemble in coveys. Such a practice would, no doubt, be esteeemed a redeeming feature in its character by many club-frequenting Benedicts of our species.

COMMON QUAIL.

COTURNIX COMMUNIS, *Bonn.*

Pl. XIX., figs. 1, 2.

Geogr. distr.—Europe generally, except the extreme north; Asia as far east as Japan; Africa as far south as the Cape and Madagascar: tolerably common and partly resident in Great Britain, breeding here and there sparingly up to the north of Scotland.

Food.—Seeds, grain, berries, leaves, and insects.

Nest.—A mere depression scratched in the ground, and lined with a small quantity of dry grass or stems of plants.

Position of nest.—In meadows or fields of growing grain.

Number of eggs.—6-14; usually 10.

Time of nidification.—VI-VII.

This species is either monogamous or polygamous. The male is said to have a distinct cry in addition to its ordinary note; this cry has been syllabled thus—" *ouen, ouen.*"

The Quail emigrates from Europe to Africa in vast flocks. It is a regular summer visitant to Great Britain, but seldom arrives earlier than May, when it chiefly confines itself to sandy soils. " Some twenty years ago," says Mr. Harting, " it used to breed regularly on the fen-lands between Newmarket and Cambridge. The nests were placed in slight scratchings or natural hollows in hay or rough grass, generally at no great distance from some marsh wall, or other commanding elevation, on which in the twilight the male might be seen disporting himself."

Mr. Howard Saunders, speaking of their distribution in England, says:—" Sparingly distributed throughout the country, there are few districts in which Quails have not at one time or another been recorded as breeding; and few also in which their appearance can be counted upon either with regularity or in anything like average numbers. In some parts of Cornwall a good many are bred, the year 1870 having proved unusually favourable for hatching; and about Bridgewater in Somersetshire a fair number nest annually. In other parts of the west they appear to be uncommon—at least, beyond Breconshire and Cheshire; but eastward they are to be found scattered about most, if not all, of the southern and midland counties."—(Yarrell's Hist. Brit. Birds, 4th ed., vol. iii., p. 124.)

Family TETRAONIDÆ.
CAPERCAILLIE.
Tetrao urogallus, *Linn.*
Pl. XVIII., fig. 3.

Geogr. distr.—From Northern to Central Europe, extending eastward far into Asia: formerly common in Great Britain, but now
comparatively rare, though probably increasing in numbers; it breeds
in the Highlands of Scotland.

Food.—Insects, seeds, berries, soft buds of pines, foliage of firs,
leaves, &c.

Nest.—A mere hole scratched in the earth by the hen bird.

Position of nest.—In the clearer parts of pine-forests in mountainous regions; under the shelter of a tree or bush.

Number of eggs.—5-15.

Time of nidification.—V-VI.

This species, which had become extinct in the British
Islands, was introduced into Taymouth in the year 1837 by
Sir Thomas Fowell Buxton; so successful was he, that in
1863 the number of birds on the estate were estimated
at 2000. " From Taymouth, the centre of restoration, and
all along the Tay Valley, as far as Dunkeld, Capercaillies
spread, and, although Perthshire still remains the head-
quarters, Forfarshire ranks not far behind. In Fifeshire,
where the woods are of smaller extent, the species is more
local, and in Kinross-shire, where there are no extensive
pine-woods, it is comparatively rare. It is merely a
straggler to Clackmannanshire, but through Stirlingshire
it is advancing, and will probably extend in time to the
southern counties of Scotland by that route."—(Howard
Saunders, in Yarrell's Hist. Brit. Birds, ed. 4., vol. iii., p. 50.)

The Capercaillie is a polygamous bird, pairing in April
and May, in which months," says Seebohm, "the males
devote themselves almost entirely to love and war. The
scene of operations is usually a favourite pine tree, conspicuous from its position on an eminence, or in an open
part of the forest, but it is said that a flat-topped rock in
the forest is sometimes chosen as the *lek-ställe*, or 'laking-
place.' Just before sunrise, and immediately after sunset,
are the times chosen by the male to repair to the ' laking
place ' which he has frequented for years." " Only the
strongest birds are allowed to ' spel,'* the younger and
weaker ones being obliged to stand aside until they have
fought their way into the privileged circle."—(Hist. Brit.
Birds, vol. ii., pp. 441, 442.)

---

* A name given to the love-song.

RED GROUSE.
Lagopus scotious, *Lath.*
Pl. XVIII., figs. 4, 5.

Geogr. distr.—Heathy districts of Great Britain: breeds in the northern counties of England, in Scotland, Wales, and many parts of Ireland.

Food.—Tender tops of heath and species of *Arbutus*, berries, and grain.

Nest.—A mere shallow cavity scratched in the ground, lined with grasses, heather twigs, or any rubbish that may chance to fall in.

Position of nest.—Usually in long heather.

Number of eggs.—7-8; if more, they are probably the product of more than one pair.

Time of nidification.—III-VI.

" The Red Grouse, which has hitherto been met with only in the British Islands, breeds on most of the high, heathy moors of this country, especially those in the north of England and the Highlands of Scotland. It is abundant also on most of the Western Islands, and is met with sparingly in Orkney, but has never been found in Shetland." —(Hewitson's Illust. Eggs Brit. Birds, vol. ii, p. 235.)

The moors on which this species breeds are, however, strictly preserved, and therefore special permission must be obtained before one can collect upon them.

Mr. Seebohm observes, " In an average year most nests will contain seven or eight eggs. Birds which breed late on the high ground do not seem to lay fewer eggs than those which breed early in the more sheltered situations. The sitting bird does not easily forsake her eggs. You may watch her daily as she sits upon them ; you may even catch her eye without frightening her away. You may send her off *cok-cok-cokking* in alarm, by accidentally almost stumbling over the nest ; and you may handle the eggs without much danger of causing her to " forsake." Game-keepers are always very anxious to impress upon trespassers the fact that it is of the utmost importance not to disturb the birds during the breeding-season. The real truth is, that if strangers were allowed on the moors at this season of the year the danger would be, not that the birds would forsake the eggs, but that the eggs would forsake the birds." —(Hist. Brit. Birds, vol. ii., p. 430.)

The Red Grouse is monogamous, and assists the hen bird to feed the young.

COMMON PTARMIGAN.
LAGOPUS MUTUS, Leach.

Pl. XVIII., fig. 6.

Geogr. distr.—Mountains of Asia and N. Europe, also the more elevated ranges of S. Europe: in Great Britain this species appears to be confined to the mountainous regions of Scotland.

Food.—Seeds, berries, buds, leaves, and ground fruit.

Nest.—A mere depression in the ground, sparingly lined with grassbents or slender twigs.

Position of nest.—At a considerable height; upon bare stony spots, or under the shelter of some small bush or large stone.

Number of eggs.—8-12.

Time of nidification.—V.

Hewitson remarks that, " although the Ptarmigan breeds in various parts of Scotland, the eggs are very difficult to obtain. Its breeding-places are those bare stony spots which cover a portion of most of the higher mountain ridges, amongst which it finds secure retreat. Its similarity of colouring is so great, and its heedlessness of danger is such—for it will remain closely crouched till you approach within the shortest distance of it—that it thus eludes discovery."—(Ill. Eggs Brit. Birds, vol. ii., p. 236.

Robert Mudie, in his 'Feathered Tribes of the British Islands' (vol i., pp. 29, 30), says:—" Usually about the month of June they disperse in pairs (they are monogamous), a little further down the mountain than their winter haunts, and there scrape a circular nest, in which the eggs, varying from six to twelve, or fifteen, the average being about ten, are deposited."

" It is, no doubt, in order to find more abundant food, and more especially to find insects for the newly-hatched young, that the Ptarmigan descends, and chooses during the breeding season a region having a temperature a little higher."

In the above account there appear to be two errors, the eggs of the Ptarmigan being laid early in May, and its food consisting, as Mr. Seebohm says, " almost exclusively of vegetable substances, such as the seeds, buds, and tender shoots of mountain-plants, especially the heath and ling. This fare is varied in autumn with berries of various kinds and ground fruit."

BLACK GROUSE.

TETRAO TETRIX, *Linn.*

Pl. XVIII., fig. 7.

Geogr. distr.—From Scandinavia to Spain and Italy, extending eastwards through Siberia to China: in Great Britain it is more numerous in the north than the south, occurring in most parts of Scotland.

Food.—Corn, berries, heath, buds, pine leaves, and insects.

Nest.—A mere depression scratched in the earth by the hen, and lined with fragments of heather, bracken and leaves.

Position of nest.—In retired spots in heather, in newly-planted ground, in hedge-rows, or under bushes.

Number of eggs.—8-10.

Time of nidification.—IV.

A polygamous species, the hens making advances in the early morning. It is easily domesticated, and has been known to breed with the barn-door fowl. As with the Capercaillie, it chooses certain localities in which to pair. These are locally known as "laking places," the meaning of the word being apparently represented in vulgar parlance by the slightly altered word "larking"; any way, here the males congregate and fight for the hens, which are attracted thither by their love-call. Dixon says that, "Throughout the laying-season the Blackcock is a noisy and pugnacious creature; and, once the full complement of eggs is deposited by the female, he quits her society probably for ever, leaving her to hatch and rear her brood unaided."

In addition to the Common Fowl, the Blackcock has been known to cross with the Capercaillie, the Red Grouse, the Willow Grouse, the Hazel Grouse, and even the Pheasant.

Mr. Howard Saunders writes: "The increase of population, the enclosure of wastes, and the drainage of boggy lands, have combined to curtail the area over which the Black Grouse formerly roamed in the south of England." He then mentions the following counties in which it is still to be found—Surrey, Berkshire, Hampshire, Sussex, Wiltshire, Dorsetshire, Somersetshire, Devon and Cornwall; also in Brecon, Radnorshire, and some other Welsh counties; in Shropshire, Staffordshire, and Nottinghamshire, "north of which they are found—although locally, and in some cases owing to introduction—in every county in England.—(Yarrell's Hist. Brit. Birds, 4th ed., vol. iii., pp. 61, 62.)

FAMILY ŒDICNEMIDÆ.

GREAT PLOVER (OR STONE CURLEW).

ŒDICNEMUS SCOLOPAX, *Gmel.*

Pl. XIX., fig. 9.

Geogr. distr.—Temperate and S. Europe in suitable localities; eastward as far as India; resident in N. Africa: in Great Britain it is local and less common than it used to be; it arrives in April, and leaves again in October.

Food.—Mollusca, worms, and insects in various stages.

Nest.—A mere depression scratched in the earth.

Position of nest.—In waste sandy flats upon the sea coast and other open desert localities.

Number of eggs.—2-3; usually 2.

Time of nidification.—V-VI; or even as late as September.

Respecting the breeding-places of this species in Great Britain, Mr. Saunders writes :—" Upon the chalk downs of Dorsetshire it is to be found breeding regularly; also, subject to the hostile influences of enclosure and cultivation, in Wiltshire; Hampshire (visiting the Isle of Wight on passage and in winter); Sussex; Kent, especially on the hills above Romney Marsh; Bedfordshire and Hertfordshire, notably on the chalk hills about Tring; and so on, through Cambridgeshire, to Suffolk and Norfolk, where it finds the conditions more congenial than anywhere else in these islands. On either side of these main lines the Stone Curlew appears to be a straggler; but it is found breeding in small numbers in Rutland and Nottingham, and the late E. Blyth obtained its young in Worcestershire. It is still found in the Wolds of Lincolnshire, and across the Humber it continues to breed, although in decreasing numbers, in a few localities in the East Riding, but to West Yorkshire it is only a straggler; and in Lancashire, Cheshire, and Wales its occurrence is very rare, if not absolutely unknown.—(Yarrell's Hist. Brit. Birds, 4th ed., vol. iii., pp. 226, 227.)

Personally I have only met with the Stone Curlew in Hampshire, where I heard it uttering its plaintive cry towards dusk whilst flying over a dreary patch of moorland. As a rule, this bird only lays once in the year, but if the first clutch is taken it will lay a second time. The egg which I have figured is in the collection of Mr. H. Dresser, and, I believe, fairly represents the species.

Family CHARADRIIDÆ.

LAPWING.

Vanellus vulgaris, *Bechst.*

Pl. XIX., figs. 4-6.

Geogr. distr.—In summer the northern and central portions of the Palæarctic region, migrating to N. Africa at the approach of winter; eastward as far as Japan: in Great Britain it is common and resident.
Food.—Insects in various stages, Mollusca and Crustacea.
Nest.—A mere cup-shaped hollow scratched in the earth.
Position of nest.—In damp places on moors or saltings, or unfrequented places in old fallow land.
Number of eggs.—4; perhaps rarely 5.
Time of nidification.—V-VI; June.

The eggs are placed in the nest with the points all together in the centre. The nest is difficult to find, as it is a mere hollow in the grass or earth, containing (to the casual observer) three large mottled stones. The bird also does its utmost to distract one's attention, wheeling round and incessantly uttering its plaintive note (which has earned it the popular name of " Pee-weet "), and even feigning to be maimed in order to try and draw one from the neighbourhood of the nest.

I have a nest obtained in Sheppy in May, 1885 ; the hollow is rather triangular than circular, measuring three and a half inches in front and four in longitudinal diameter ; its depth in the centre is nearly two inches; it contains three eggs.

The eggs of this species, being considered a great delicacy, are daily sought for by old men, women, and children, who either sell them on the spot for two shillings a dozen, or send them up to the London market, where they fetch more than double that price; many other eggs are, however, disposed of as Lapwing's eggs, and, amongst those offered for sale, I have seen not only those of Gulls and Terns, but even eggs of the Moorhen.

In seeking for the eggs, little notice is taken of the male bird, which flies overhead trying to distract attention, whilst the hen bird runs from the nest for a short distance, and then rises quietly, still, however, keeping tolerably near to the ground; if, therefore, the spot where the hen begins to run is carefully noted, there is every likelihood of finding the nest.

K

GOLDEN PLOVER.

CHARADRIUS PLUVIALIS, L*inn*.

Pl. XIX., figs. 7-8.

Geogr. distr.—Europe extending eastward as far as the Yenesay and southward to Belgium, Holland, and N. Germany, wintering on the shores of the Mediterranean; Asia as far east as Beloochistan; Africa as far south as the Cape : in Great Britain it breeds chiefly in Scotland and Ireland, but also in the extreme south-west of England and some parts of Wales.

Food.—Seeds of weeds, berries, insects, slugs, small sea Mollusca, and worms.

Nest.—A mere depression in the ground lined with dry grasses, fragments of heather, and moss.

Position of nest.—On heath-covered moors and swampy ground.

Number of eggs.—4.

Time of nidification.—V-VI ; middle of May.

Mr. Howard Saunders writes : " The Golden Plover is found during summer breeding on the high hills and swampy grounds of Great Britain and Ireland. In England it is believed to breed sparingly in Devonshire, and perhaps in Somerset, and it is known to do so in Breconshire, and some other counties of Wales and its borders. From Derbyshire onwards it becomes more abundant as a nesting species, and in Scotland it is generally distributed, being especially numerous in Sutherlandshire. It is a familiar bird on the moors of the Orkney and Shetland Islands, and in the Hebrides the numbers which descend to the sandy pastures and shores are said by Macgillivray to be astonishing. Throughout Ireland it is to be found breeding in suitable localities."—(Yarrell's Hist. Brit. Birds, 4th ed., vol. iii., p. 273.)

According to Hewitson, " it is a very watchful bird, and usually discovers itself long before you approach it, by its clear and plaintive whistle, which may be heard at a great distance, and is very deceptive." He continues, " Though, as I have just stated, usually very wary and difficult to approach during the earlier days of incubation, it will sometimes, when the eggs are nearly hatching, almost allow itself to be trodden upon before it leaves its nest." —(Ill. Eggs Brit. Birds, vol. ii., p. 249.)

The food of the Golden Plover in summer consists, according to Seebohm, principally of worms and insects, and upon this diet the young begin to feed as soon as hatched. One brood only is reared in a season.

DOTTEREL.

EUDROMIAS MORINELLUS, *Linn.*

Pl. XX., fig. 1.

Geogr. distr.—Western Palæarctic region, not occurring in the Oriental region; winters in Africa north of the Equator: in Great Britain it occurs more commonly in the eastern than the western counties of England, but its favourite haunts are the wilder parts of Scotland; in Ireland and Wales it is rarer.

Food.—Young shoots of alpine plants, insects in various stages, and worms.

Nest.—A mere depression in the ground.

Position of nest.—Usually near to, or on the summits of high mountains amongst moss and other vegetation, and generally near to a fragment of rock or stone.

Number of eggs.—3.

Time of nidification.—VI-VII.

This is a summer visitor to Great Britain, arriving in April or May, and leaving again in September; it is supposed still to breed (as formerly), though sparingly, in the mountains of the English lake district, in Perth and the north of Scotland, and in the Orkneys.

Mr. Seebohm says that " the eggs of the Dotterel vary in ground-colour from greyish buff to ochraceous buff, with sometimes the faintest possible tinge of olive, and are blotched and spotted with rich dark brown and with underlying markings of inky grey. The surface markings are generally large, concealing a large portion of the ground colour, and are often confluent, especially on the larger end of the egg." " The eggs vary considerably in shape, some being almost as pointed at the large end as at the small, whilst others are pear-shaped.' '" The only eggs of a British bird at all likely to be confused with those of the Dotterel are certain varieties of those of the Arctic Tern, some of which are almost indistinguishable from those of the Dotterel, but the latter have fewer and smaller underlying markings."—(Hist. Brit. Birds, vol. iii., p. 33.)

According to Mr. Heysham's account, which has been copied into almost every book on British Birds, the eggs of this species may be found fresh both in June and July.

The egg which I have figured was selected for that purpose by Mr. H. Dresser from his series, as being typical.

RINGED PLOVER.

ÆGIALITIS ASIATICA, *Pallas*.

Pl. XX., fig. 2.

Geogr. distr.—Europe generally; western Asia; Africa as far southward as the Cape; also recorded from Australia: in Great Britain generally distributed and partially resident.

Food.—Insects, spiders, Crustacea, Mollusca, and sea worms.

Nest.—Sometimes none, but generally a mere depression amongst tufts of long grass, or scratched in shingle, with a few shells in the centre.

Position of nest.—Frequently just beyond the reach of the water upon little hillocks of sand upon the sea-beach; also occasionally near inland pieces of water.

Number of eggs.—4.

Time of nidification.—IV-VIII; April.

When disturbed the Ringed Plover rarely takes long flights, but soon alights and runs rapidly on the sand. Its note is a clear, shrill whistle. It is very fond of barren, shingly beaches, and is essentially a shore bird. It may also be seen by the sides of large rivers, of inland lakes, or large ponds.

The Ringed Plover has been known to lay as early as the 23rd March, but the middle of April to the middle of May is the best time at which to commence looking for the eggs.

Hewitson says, "It makes no nest, but lays its four conical eggs in a slight hole on the surface of the ground, either amongst small gravel or upon the little hillocks of sand which occur so commonly on our flat sea-beach. In some of these substitutes for a nest, which I have seen, the eggs presented a very beautiful appearance upon the clean white sand; frequently near the root of some tall grass, which waved over them as a protection from the storm. During the breeding season the Ring Dotterel is ever on the alert and on wing long before you reach its eggs, making its circuits round you, and uttering its sweet plaintive whistle of alarm—a sure indication that you are in the near neighbourhood of its eggs or young ones."—(Ill. Eggs Brit. Birds, vol. ii., p. 255.)

It is believed that the Ringed Plover rears but one brood in a season.

KENTISH PLOVER.

ÆGIALITIS CANTIANA, *Lath.*

Geogr. distr.—Central and Southern Europe ; Asia as far eastward as Japan ; Africa as far southward as the Cape ; common in N. Africa : in Great Britain it has occurred sparingly in Norfolk ; it is chiefly found, however, along the southern coast of England.

Food.—Insects in all stages, worms, Mollusca, and Crustacea.

Nest.—A mere depression in the ground, sand, shingle, or earth, usually without lining, rarely lined with fine rootlets and a few grass bents.

Position of nest.—Often in a bunch of wild oats or willow-shoots on overgrown downs thickly covered with grass, rushes, brambles, &c. ; but usually on bare sandy flats, or in shingle or sea-weed cast up by the waves.

Number of eggs.—3 ; rarely 4.

Time of nidification.—V.

Of this species Mr. Seebohm writes as follows :—" It is one of the most local of British birds, and has only occurred very sparingly on the south and east coasts of England, as far north as Flamborough Head in Yorkshire, and as far west as Cornwall. Its only breeding-places in this country appear to be on the coasts of Kent and Sussex ; but even there it is a rare bird, and is fast disappearing before the inroads of collectors."

" Although the bird does not exactly breed in colonies, numbers of its nests are made close together ; and even where the bird is not common all the birds of a district often rear their young on the same suitable patch of coast."

" The Kentish Plover does not sit very close. The instant danger threatens, the watchful male conveys the alarm to his brooding mate, and she leaves her eggs at once, conscious that their protective colouring will ensure their safety. Only one brood appears to be reared in the year.—(Hist. Brit. Birds, vol. iii, pp. 25, 26, 27.)

This species appears to breed freely in the island of Guernsey and the neighbouring islets.

The egg which I have figured is from the collection of Mr. H. Dresser.

TURNSTONE.

STREPSILAS INTERPRES, *Linn.*

Pl. XX., fig. 4.

Geogr. distr.—Almost cosmopolitan : common, and, in a sense, partly resident in Great Britain, being most numerous on the southern coasts.

Food.—Insects, Crustacea, Mollusca, and worms.

Nest.—A mere depression in the earth, lined with a few grass-bents.

Position of nest.—On sandy or rocky soil, in bare sandy places or flats covered with heath and a few stunted junipers.

Number of eggs.—4.

Time of nidification.—VI.

Although the breeding of this species in Great Britain has not as yet been distinctly proved, there is every probability that it does so. In the third volume of Yarrell Mr. Howard Saunders, after mentioning the fact that "birds in breeding-plumage have frequently been observed on our coasts, sometimes in pairs, all through the summer," states that on the 28th May, 1861, a pair rose from a most suitable locality on Lundy Island, but that the male, unfortunately, was shot by Mr. Saunders's companion ; that Mr. T. E. Buckley has seen the bird on the west coast of Harris in July, and believes that it breeds there ; that the late Dr. Saxby saw a Turnstone on Unst in the Shetlands, on the 16th June, and found three eggs which he supposed to belong to it ; and lastly that he himself, in July, 1879, saw a pair on a neighbouring island.

Mr. Seebohm says :—" When we bear in mind how little is known of the ornithology of the islands on the wild west and north of Scotland, and remember that it breeds at no great distance from Copenhagen, it is difficult to believe that the Turnstone does not breed on the Hebrides in limited numbers."—(Hist. Brit. Birds, vol. iii., p. 12.)

This species almost always remains near water, either on the sea-shore, the margins of lakes, or at the mouths of large rivers, and picks up a living amongst the stones, sea-weed, or mud, small thin-shelled Mollusca and Crustacea appearing to be its favourite food.

The egg which I have figured is in the collection of Mr. Dresser. Some eggs are more olive-green in tint, and heavily blotched with rufous-brown.

OYSTERCATCHER.

HÆMATOPUS OSTRALEGUS, *Linn.*

Pl. XX., figs. 5, 6.

Geogr. distr.—Almost the entire Palæarctic region; generally distributed in Europe; Asia as far south as Ceylon; in the winter ranging far down the African coast: occurs in almost all parts of the British coast, especially in the north.

Food.—Mollusca, Crustacea, and other marine animals; when domesticated, fish refuse and insects of all kinds.

Nest.—A mere depression, lined with a few straws or fragments of plants, shells, gravel, or stones.

Position of nest.—Amongst shingle or gravel, just above high-water mark; in short, stunted thrift, or on a tussock of sea-pink.

Number of eggs.—3-4; usually 3.

Time of nidification.—IV-VI; end of May or June.

This bird is often tamed and turned loose in a garden, where it is very useful in destroying worms, woodlice, earwigs, and other pests, with which it will gorge itself until unable to look at another without abhorrence. It has been known to live in confinement for thirty years, and therefore the good which it is capable of is incalculable. It is moreover not only an ornament to a garden, but a great source of amusement, even its manner of tripping along being ludicrous. An aunt of my wife has one of these birds in company with a Gull in her garden, and the strong-billed bird generally keeps the Oystercatcher at a respectful distance until it has selected the choicest morsels, but sometimes the latter approaches stealthily and scampers off with a dainty fragment whilst the Gull is occupied.

Mr. Howard Saunders says:—"Although principally found on or near the coast, it is a mistake to suppose that the Oystercatcher does not straggle inland, for examples have been killed even in the Midland Counties. In Scotland many pairs breed on the Don, the Tay, the Spey, the Findhorn, and on some island lochs twenty or thirty miles from the sea."—(Yarrell's Hist. Brit. Birds, 4th ed., vol. iii., p. 295.)

Oystercatchers are said to be always more or less gregarious in their habits, and to be very difficult to satisfy in the preparation of their nest, several being completed and deserted, all apparently as good as the one in which they lay.

Family SCOLOPACIDÆ.

COMMON REDSHANK.

Totanus calidris, *Linn.*

Pl. XX., fig. 7.

Geogr. distr.—Europe generally; Asia as far as S.E. Siberia and China; in the winter it is found in Africa as far south as the Cape: in Great Britain it is resident, being most common in Scotland.

Food.—Insects, Crustacea, Mollusca, and worms.

Nest.—A mere depression in the ground, or a hollow trodden down in the centre of a tuft of grass, lined with a few dry grasses, straws, or fragments of moss, heather, or reed.

Position of nest.—On wet heathy moors or marshy wastes, sometimes under a tall weed or heather bush.

Number of eggs.—4.

Time of nidification.—IV-V.

This bird is by no means uncommon in England, even in the summer months, though much less so than formerly, owing to the draining of many marshes and reclaiming of damp wastes, upon which it used to abound. Though pretty generally distributed wherever the country is suitable, it has now become, of necessity, somewhat local; in Scotland and the islands to the west and north of that country it is abundant.

During the nesting season the Redshank is sociable, many nests being prepared within a small area of ground. When disturbed from the nest the birds are said to be very clamorous, flying round the intruder, and making an incessant, shrill, piping note. Like other species of a similar nature, the number of eggs in each nest is invariably four, and those constantly placed with their smaller ends in the centre.—(Newman, in Montagu's Diet. of Brit. Birds, p. 269.)

As Hewitson observes, the eggs are " much like those of the Peewit in size and general appearance;" they are, however, paler, less richly coloured, have, as a rule, smaller markings, and are more pointed at the small end.

Only one brood is reared in a season.

GREENSHANK.
TOTANUS CANESCENS, *Gmelin.*
Pl. XX., fig. 8.

Geogr. distr.—Entire Palæarctic region in summer, but breeding principally in the north; S. Africa and Malay Archipelago to Australia in winter; also said to visit S. America: in Great Britain it breeds in Scotland.

Food.—Mollusca, Crustacea, and fish.

Nest.—A shallow depression in a dry tuft of grass, lined sparingly with dry, wiry grass and fragments of heath.

Position of nest.—In or on the borders of marshy ground, by the side of a loch, or in open forest, often at some distance from water.

Number of eggs.—4.

Time of nidification.—V.

Although this species is tolerably common in Ireland, in autumn and spring, and many localities are believed to be suitable for breeding in, no instances of its nesting there are recorded. In Scotland and the Hebrides, however, its breeding is well established, and is said to be on the increase. Nests were also discovered by Saxby in Shetland. When on migration the Greenshank has occurred in almost every county of Great Britain, but more especially the eastern counties. It is not, however, known to breed farther to the south than Perth and Argyle. It arrives on our coasts in April and May, and leaves us again in September, though a few remain with us at times, especially in Ireland, during the winter. In habits it is said much to resemble the Redshank, though in its nesting it does not exhibit the same sociability. It rears but one brood in the year.

Mr. Seebohm says:—"The summer haunts of the Green-shank are on the mountain-heaths, on the broad moors which are studded with lochs and interlaced with streams and pools. On its arrival in this country it frequents the coast for a short time, showing preference for shores that are low and muddy; but it soon leaves for its summer quarters amongst the hills. Both in spring and autumn it sometimes congregates into small flocks; but it is more often seen in pairs."—(Hist. Brit. Birds, vol. iii., p. 150.)

RUFF (FEMALE, REEVE).

TOTANUS PUGNAX, *Briss.*

Geogr. distr.—A western Palæarctic species, breeding as far north as land extends, as far south as the valley of the Danube and the Kirghis Steppes, as far east as the Taimur peninsular and West Dauria, reaching and probably breeding in the upper valley of the Amoor; it winters in suitable places throughout Africa, N. India, and Burma; has occurred in Ceylon, Yezo (Japan), Maine, Massachusetts, New York, and Spanish Guiana.—(Seebohm, Hist. Brit. Birds.)

Food.—Insects in various stages, worms, sand-hoppers, and seeds.

Nest.—A mere depression lined with dead grass and sedge.

Position of nest.—On the ground in a swamp, placed in the middle of a tussock of coarse grass.

Number of eggs.—4.

Time of nidification.—V.

This species, which was formerly a regular breeder in Great Britain, has now almost, if not altogether, ceased to do so; Mr. Howard Saunders, indeed, says that, "In Norfolk it is possible that a pair or two may still nest, and in Lincolnshire a nest containing two eggs was taken in 1882." Mr. Seebohm speaks of "a few pairs still occasionally breeding in the Norfolk broads." Looking to the fact that the species is not becoming commoner as an indigenous breeder, and the little likelihood of its eggs being taken by any reader of this volume, I concluded that it would be next to useless to figure it.

The eggs, according to Seebohm, vary from an almost neutral pale grey to pale greenish grey, the overlying spots reddish brown, and the underlying spots pale greyish brown; Hewitson's figures represent them of about the size and form of my egg of the Redshank, but paler, the markings having a smeared appearance, and the ground tint varying from pale greyish green to whity brown.

As with the Capercaillie, the Ruffs fight for supremacy, and the place set apart for their battles is known as a "hill." Though their aspect when fighting is very ferocious, they appear to do one another little mischief, the conquered Ruff being always more frightened than hurt.

COMMON SANDPIPER.

TOTANUS HYPOLEUCUS, *Linn.*

Pl. XX., fig. 10.

Geogr. distr.—Europe generally; Asia to the Malay Archipelago; Africa to the Cape; Australia: in Great Britain it breeds regularly in Somerset, Devon, Wales, Yorkshire, Scotland, Shetland, and Ireland, and occasionally in Cornwall and Sussex.

Food.—Insects, Mollusca, Crustacea, and sea-weed.

Nest.—A mere hole lined with dry grass and moss, or without any lining.

Position of nest.—Amongst herbage upon banks of rivers and streams, in gravel beds amongst pebbles, or in irregularities upon the surface of a bare rock.

Number of eggs.—4.

Time of nidification.—V.

It is said that this bird often betrays the proximity of its nest by its anxiety, in which respect it is by no means peculiar. On the banks of the Findhorn, in Elgin, Scotland, it is stated to nest amongst whins and alders. According to Hewitson, it "frequents almost every river, skimming over the surface, and uttering its sweet melancholy whistle. It lays its eggs either amongst the large dockens that grow upon the banks, or upon the beds of gravel by the margins of the stream. In the former situation, where there is apparently less need, it makes a slight nest by collecting a little dry grass, and placing it in a hole scratched for that purpose; in the latter none, contenting itself by placing its eggs in a slight depression amongst the gravel. Here it is, however, by no means easy to discover them, placed as they are amongst the small pebbles. I have found them upon the bare flat rock, where nothing but a very slight inequality in the surface kept them in their places."—(Ill. Eggs Brit. Birds, vol. ii., p. 293).

Mr. Howard Saunders says that it "makes a slight nest of moss and dried leaves in a hole on a bank near fresh water, generally under shelter of a bunch of rushes or a tuft of grass, and sometimes in a corn-field, if it happens to extend near enough towards the water."—(Yarrell's Hist. Brit. Birds, 4th ed., vol. iii., p. 449.)

RED-NECKED PHALAROPE.
PHALAROPUS HYPERBOREUS, *Linn.*
Pl. XX., fig. 9.

Geogr. distr.—In summer in the northern portions of the Palæarctic and Nearctic regions, migrating southwards at the approach of winter as far as Algeria, and in Asia as far as the Aru Islands : it breeds in the extreme north of Great Britain.

Food.—Marine insects, Crustacea, Mollusca, and worms.

Nest.—Cup-shaped and deep ; of dry grass and fragments of reed.

Position of nest.—In tufts of grass at the water's edge, or amongst rubbish left dry by the gradual sinking of the water in continuous fine weather.

Number of eggs.—3-4.

Time of nidification.—V-VI.

This species used to breed in the Orkney Islands, and still continues to do so in the Hebrides and Shetland, but very sparingly, and it is believed by some naturalists still to breed here and there in the counties of Perth and Inverness.* The nest is said to be about the size of that of the Titlark.

The following account of the nesting is from Mr. Seebohm's valuable work, from which (though with the free sanction of the author) I have culled so much valuable information that I feel almost ashamed of myself :—

" The Red-necked Phalarope breeds on the tundras above the limit of forest growth, and prefers marshy ground covered with long grass, similar to that frequented by the Reeves. In this long grass it builds its nest, which is a somewhat slight structure of dry stalks, generally placed in the middle of a thick tuft, so that it is not unfrequently a foot or more from the ground. In some places Harvie-Brown and I found the nests of this bird where the grass was short, and in these situations it was scarcely more than a hollow in the ground lined with dead grass. We invariably found the eggs with the small ends pointed inwards, and there was always a lining to the nest."—(Hist. Brit. Birds, vol. iii., p. 90.)

The egg which I have figured is in the collection of Mr. Dresser, but I believe I have somewhat exaggerated the green ground-colour, which is usually paler, and varies to pale or deep ochraceous in tint.

---

* Seebohm, however, evidently thinks that it has ceased to breed in these counties.

DUNLIN.

TRINGA ALPINA, *Linn.*

Pl. XXI., fig. 1.

Geogr. distr.—Northern and Central Europe generally; eastward in Asia as far as China and Japan; America: in the winter it occurs in India, N. Africa, and southern N. America: in Great Britain it breeds in some parts of England, in Scotland and the Scotch islands.

Food.—Insects, Crustacea, Mollusca, and marine worms.

Nest.—A depression in the ground lined with fibrous roots, a little dry grass, or fragments of heath.

Position of nest.—Usually near the sea, amongst short grass or moss on moors or in open swampy flats.

Number of eggs.—4.

Time of nidification.—V-VI; June.

Hewitson says that, "the eggs of the Dunlin are usually placed very snugly either amongst heath or under a tuft of long dry grass, and are then difficult to find." "The Dunlin can scarcely be said to make a nest, for the most part merely rounding into form the grass or moss amongst which it is about to lay its eggs. Sometimes pieces of heath and a little dry grass are added, but this is not often." He adds, "I once found the nest of this species upon one of the unfrequented moors of Shetland, and not then well knowing the eggs, I left them till I could return with my gun to secure one of the birds. I did return a few hours afterwards, but the eggs were gone; and, though I have no evidence to prove it, I have myself no doubt that the birds had removed them to a place of safety."— (Ill. Eggs Brit. Birds, vol. ii., pp. 313, 314).

In England, during the breeding season, this bird is both rare and local, but a few pairs probably remain to nest in Cornwall, Devon, Cheshire, perhaps Lincolnshire, Yorkshire, and Lancashire. In Cumberland it is more numerous, as also in various counties of Scotland, such as Lanarkshire, Renfrewshire, the western counties, and the Scotch islands, being by no means rare in the Orkneys and Shetlands. In Ireland, though many localities appear suitable, a comparatively small number appear to breed.

The nest, being usually concealed under a tuft of grass, is by no means easy to find unless the bird is flushed from it.

COMMON SNIPE.

GALLINAGO CŒLESTIS, *Frenz.*

Pl. XXI., fig. 2.

Geogr. distr.—Europe and Asia; ranging into Northern Africa, the islands of the Bay of Bengal and the Philippines. Resident in Great Britain, and to be met with in all suitable localities; more abundant in Ireland than in Scotland or England, on account of the nature of the country.

Food.—Insects, worms and Mollusca.

Nest.—A mere rounded depression in the ground, lined with dried grasses or other herbage.

Position of nest.—Amongst grass or rushes.

Number of eggs.—4.

Time of nidification.—IV-VII.

The Snipe is far less abundant in winter than in spring and autumn. It occurs in low woods, damp pastures, and marshy places, either singly, in couples, or small companies. Its usual cry is a succession of two or three tremulous piping notes, but in the breeding season it makes a sound something like the hum of a bee.

Mr. Howard Saunders says that "the feeding-ground of the Snipe is by the sides of land springs, or in water meadows; and in low flat countries they are frequently found among wet turnips. The holes made with their bills, when searching for food, are easily traced. The end of the beak of the Snipe, when the bird is alive, or if recently killed, is smooth, soft, and pulpy, indicating great sensibility. If the upper mandible be macerated in water for a few days, the skin, or cuticle, may be readily peeled off. The external surface presents numerous elongated, hexagonal cells, which afford at the same time protection to, and space for the expansion of, minute portions of nerves supplied to them by two branches of the fifth pair; and the end of the bill becomes, in consequence of this provision, a delicate organ of touch, to assist these birds when boring for their food in soft ground."—(Yarrell's Hist. Brit. Birds, 4th ed., vol. iii., pp. 346, 347.)

The Snipe is generally distributed throughout Great Britain, breeding in all marshy localities, though less abundantly in England than in either Scotland or Ireland. The nest is usually placed in a clump of sedge, coarse grass, or rushes. When flushed from its eggs this bird flies off without uttering any cry.

WOODCOCK.
SCOLOPAX RUSTICULA, Linn.
Pl. XXI., fig. 3.

Geogr. distr.—Europe generally, breeding in the northern and central countries and wintering in the south; in Asia eastwards to Japan and southwards to Ceylon; N. Africa: in Great Britain it breeds almost everywhere, but not in any numbers.
Food.—Worms and insects in various stages.
Nest.—A mere depression.
Position in which eggs are deposited.—Amongst dead leaves or dry grass in open places in woods.
Number of eggs, 4.
Time of nidification.—III.

This species usually travels by night, and its flight is not unlike that of an owl, but when flying high it is said to move swiftly. Its journeys are taken either singly or in couples. Its presence in a wood is discovered by searching for a moist spot or spring round about which the large white excreta are at once noticeable. Its cry is hoarse, and has been likened to the croak of a frog.

Mr. Harting informs me that the Woodcock breeds in England much more commonly than most persons suppose. Mr. T. Monk, of Lewes, some years since, was at considerable pains to obtain statistics as to the number of Woodcocks remaining to breed in the eastern division of Sussex; and, extraordinary as it may appear, the conclusion he arrived at was to the effect that in seven districts of East Sussex, comprising twenty-one parishes, there were annually, on an average, from 150 to 200 nests of this bird. When flying over the woods just after dark the Woodcock is said to utter a shrill piping cry, although when flushed during the day-time it is mute.

The eggs are laid in a warm and sheltered place upon dead oak or fern leaves, which, by their mere weight, they depress into a hollow. To blow a Woodcock's egg with a small hole is, as I know to my cost, a task requiring both patience and perseverance, as the white is very gelatinous.

BLACK-TAILED GODWIT.

LIMOSA ÆGOCEPHALA, Linn.

Pl. XXI., fig. 4.

Geogr. distr.—Europe, through Siberia and India, to China and Japan; also N. Australia: it is doubtful whether this bird still breeds with us; formerly it was known to breed in Lincolnshire, Norfolk, Cambridgeshire, and Huntingdonshire.

Food.—Insects, worms, Mollusca.

Nest.—Composed of dried grass and other herbage.

Position of nest.—Concealed amongst coarse vegetation in low-lying meadows and marshy land.

Number of eggs.—4.

Time of nidification.—IV-V usually May.

This species appears to have become extinct as a breeder in Great Britain. Mr. Howard Saunders says :—" The Black-tailed Godwit was accustomed to resort to the marshes of Norfolk and the fens of the Isle of Ely and of Lincolnshire, down to about the year 1829, by which time the drainage of suitable haunts and the persecution of gunners, netters, and egg-gatherers, had done their work. A few pairs appear to have nested irregularly until a later date, for Mr. E. S. Preston is said to have obtained three eggs, which were stated on good evidence to have been taken near Reedham in Norfolk in 1847. A few birds now linger for a few days in spring about the localities where their predecessors found suitable breeding-grounds, but they pass on, and at the present day the Black-tailed Godwit is only known as a visitor on migration."—(Yarrell's Hist. Brit. Birds, 4th ser., vol. iii. p. 490.)

Writing of this bird in 1846, Hewitson stated that it bred occasionally, though sparingly, in the fens of Cambridgeshire and some of the marshy districts of Norfolk. One of the eggs which he figures is coloured like my egg of the Whimbrel, but is rather smaller and mottled, and spotted more like that of the Curlew. He mentions, however, that some varieties have scarcely any perceptible markings. My figure, taken from an egg presented to me by my valued friend the Rev. W. Bree, has, unfortunately, been printed of too lively a green, the true tint being duller and more olivaceous.

WHIMBREL.
NUMENIUS PHÆOPUS, *Linn.*

Pl. XXI., fig. 5.

Geogr. distr.—Entire Palæarctic region; Indo-Malaysia; Africa to
the Cape: in Great Britain during the summer it occurs far to the
north, breeding in Scotland, especially in the Orkney and Shetland
Islands.
Food.—Worms, Mollusca and Crustacea.
Nest.—A mere depression in the soil lined with a few leaves and
grass-stalks, or without lining.
Position of nest.—On a slight elevation in marshy ground where
the soil is peaty and mossy.
Number of eggs.—4.
Time of nidification.—V-VI.

Captain Henry W. Feilden, in the 'Zoologist' for 1872,
mentions having taken a nest of this bird with four eggs
on the 16th June in a singular position, in the Faroe
Islands; the nest was placed near a rill, between two
blocks of stone which only just allowed sufficient space for
the bird to squeeze between them. He states that, when
breeding, the Whimbrel is pugnacious, and constantly on
the alert to drive away intruders from the vicinity of its
nest; and that he has watched it by the hour chasing the
Lesser Black-backed Gull. When engaged in these combats
its flight is rapid and arrow-like, and it constantly repeats
its trilling cry, which has not inaptly been described as
resembling the words "tetty, tetty, tetty, tet" quickly
repeated. (*See* p. 8248).

Mr. Seebohm says that "the favourite breeding-grounds
of the Whimbrel are the moors and heaths in close
proximity to the sea. When the vicinity of their nest is
intruded upon the Whimbrels fly into the air and circle
round and round. The nest is very slight, a little hollow
amongst the heath, or under the shelter of a tuft of coarse
grass in a dry part of the swamp, and is lined with a few
scraps of dry herbage. The eggs are usually laid at the
end of May, and from that date they may be obtained until
the end of June."—(Hist. Brit. Birds, vol. iii., p. 102.)

I have a very badly-blown egg of this bird, said to have
been taken in Mid-Kent; but this is most improbable.

L

COMMON CURLEW.
NUMENIUS ARQUATUS, *Linn.*
Pl. XXI., fig. 6.

Geogr. distr.—Throughout Europe and Asia down to South Africa; also found in the islands of the Indian Archipelago. Always to be met with in Great Britain, and therefore partially resident.
Food.—Insects, worms, Crustacea, Mollusca, and fish.
Nest.—A mere grassy tussock.
Position of nest.—On level marshy or undulating moors and sheep pastures near water.
Number of eggs.—4.
Time of nidification.—IV-V.

In his 'Sketches of Bird Life,' Mr. Harting says that the Curlew depends more upon the sense of sight than that of hearing to save itself from its enemies, and therefore it is sometimes attracted within shooting distance by the use of a trained red-coloured dog, as nearly as possible resembling a fox. The birds will frequently chase such a dog, which gradually draws nearer to its master, and thus enables him to obtain the desired shot. The note of the Curlew is, Mr. Harting says, the loudest of any of our grallatorial birds; its usual cry is "*cou-r-lieu, cour-lieu,*" and it also has a cry resembling "*wha-up,*" which has earned it, in Scotland, the popular name of Whaup; but when pairing its note is softer—"*whee-ou, whee-ou.*"

On the sandy parts of the coast this bird feeds largely on cockles, and has been known to swallow the shell entire.

In the 'Zoologist' for 1882, p. 216, Mr. E. Cambridge Phillips says :—" I once saw a Curlew make a very determined attack on an old Carrion Crow that was probably on the look-out for one of its young ones. The Crow stood no chance against the Curlew with its grand free flight, and was soon beaten off and pursued until both were close to me."

According to Hewitson, the nest of this species, when there is any, consists of a few pieces of dried grass collected together in a hollow in some tuft of the same material."— (Ill. Eggs. Brit. Birds, vol. ii., p. 285.)

Family RALLIDÆ.
WATER RAIL.
Rallus aquaticus, *Linn.*
Pl. XXI., fig. 8.

Geogr. distr.—Throughout Europe, India, N. Africa: generally in Great Britain, and resident.

Food.—Aquatic insects, Mollusca, and worms ; vegetable diet only when pressed by hunger.

Nest.—Large and loosely constructed, of stems and bent leaves of flags, rush or sedge, and dried leaves of aquatic plants, and lined with dry fragments of reed stems.

Position of nest.—Well concealed amongst reeds and rushes in marshy ground, or upon half-floating rushes.

Number of eggs.—8-10, and sometimes more.

Time of nidification.—IV-VII.

Common in most stagnant waters, very quick in its motions ; when alarmed it is very clever in counterfeiting death, and Mr. Phillips (' Zoologist,' 1882, p. 218) mentions that once, when shooting, he flushed a Water Rail, at which he fired, when the bird immediately fell to the ground. On picking it up, it lay in his hand for some minutes motionless, and, to all appearance, dead. He was looking for a shot mark, when, chancing to turn his head away for a moment, it flew off without the slightest warning.

Though a fairly common species, the Water Rail is so shy and retiring in its habits, dwelling in the cover of the reeds and coarse vegetation of our marshes, and rarely showing, unless in danger of capture by a dog, that it appears to be quite a rare bird. In some localities, indeed, it is scarce enough, and in Shetland it is said to be by no means common ; but on the Norfolk Broads during the nesting season one may frequently hear its loud cry. The nest is difficult to find, being placed in the midst of densely-growing reeds and rank herbage. Mr. Seebohm thus describes one taken by himself :—" The nest was admirably concealed, and with all our care we only caught a momentary glimpse of the bird as she disappeared. Such a nest can only be found by accident. The perfect silence of the bird, the quiet way in which she slips off the nest and threads her way amongst the sedge and reeds, and the absolute concealment of the nest itself, which cannot be seen until the vegetation which hangs over it is pulled aside, make it an almost hopeless task to try and find a nest in such extensive reed-beds."—(Hist. Brit. Birds, vol. ii., p. 554.)

The egg which I have figured is in Mr. Seebohm's collection.

LAND RAIL.

CREX PRATENSIS, *Bechst.*

Pl. XXII., figs. 3, 4.

Geogr. distr.—Europe generally and Western Asia; extending nearly to the Arctic Circle in summer, migrating in the autumn to Africa as far southward as the Cape: generally in Great Britain in the summer, breeding in any suitable locality: it arrives towards the end of April or beginning of May, and leaves us in September or October.

Food.—Insects, especially small Coleoptera, Lepidoptera, Diptera also spiders and worms.

Nest.—A mere depressson in the ground lined with a few straws.

Position of nest.—Usually in a corn or clover field, or meadow.

Number of eggs.—8-12.

Time of nidification.—V-VI; June.

The Land Rail, or Corn Crake, though a summer visitant in our islands, may yet occasionally be seen on bright winter days, and more especially in Ireland. Mr. Harting, in his 'Summer Migrants,' suggests that these winter birds may be individuals of a late brood hybernating, that is to say in a semi-torpid condition during the dull weather. In May and June one may constantly hear the " craking " of this bird—a cry resembling the rasping of a file along a piece of steel, but rasped in one direction and pulled right off at the end " *grrrrrrrr-kin*," and the bird has such ventriloquial power * that it is impossible to tell from what direction the sound comes, sometimes sounding half-a-mile to the east and the next minute as far to the west. The note is frequently duplicated.†

As is the case with the Water Rail, the Corn Crake, when suddenly surprised and captured, simulates death with admirable fidelity, and, according to Mr. M. A. Mathew (Zoologist, 1872, p. 3316), it will take to the water and swim, after the manner of a Moorhen, taking to flight when it supposes itself to be at a safe distance. Unless compelled, however, it rarely flies, and, even when startled into taking wing, it flutters heavily just above the ground, and takes shelter at the first opportunity.

---

* This, however, is disputed by some, but so it certainly appears to me.

† The note has been likened to the word " *crake*," or " *crek*," but there is always a metallic catch at the end; possibly the sharpness of this note may render it inaudible to some ears.

BAILLON'S CRAKE.
PORZANA BAILLONI, *Vieill.*
Pl. XXII., fig. 5.

Geogr. distr.—Central and Southern Europe; Africa as far south-ward as the Cape and Madagascar; it breeds in Turkestan and winters in Persia: several instances of its breeding in Norfolk and Cambridgeshire have been recorded, but it is decidedly rare in Great Britain.

Food.—Mollusca, worms, and insects.

Nest.—Carelessly, though not loosely, formed of rush and weed, lined with fine leaves of aquatic plants and dried grasses.

Position of nest.—Upon the ground in marshy places, a little above the water's edge, in wild rush or water grass, well concealed in a bunch of sedge, the points of which are bent over so as to form a basket-like cavity.

Number of eggs.—5-8; usually 6.

Time of nidification.—V-VIII; June.

Mr. Seebohm says :—"It is not improbable that Baillon's Crake breeds in our islands every year. It is such a skulking species, and so small, that it is very easily overlooked; and the discovery of its nest and eggs some years ago was due almost entirely to accident. The first recorded instance of Baillon's Crake nesting in this country is that of Mr. A. F. Sealy ('Zoologist,' 1859, p. 6329), who described two nests that were found in the fens of Cambridgeshire. One of these was discovered on the 6th of June of that year, containing six eggs; the other, on which the female was captured, was found in the first week in August, and con- tained seven eggs considerably incubated."—Hist. Brit. Birds, vol. ii., p. 544.)

The eggs taken in Norfolk were obtained in June and July, 1866, on Heigham Sounds, near Hickling.

Mr. Howard Saunders says: "Baillon's Crake appears to be less partial to meres and open water than the Little Crake; on the contrary, it frequents the smaller marshes and swamps, especially where there is a surrounding of tamarisk and other bushes. Evening and daybreak are almost the only times when it is to be seen, unless very much pressed by a dog, and even then it is loth to take wing. The nest, concealed amongst the aquatic vegetation, is composed of dry flags and sedge."—(Yarrell's Hist. Brit. Birds, vol. iii., p. 157.)

SPOTTED CRAKE.

PORZANA MARUETTA, *Leach*.

Pl. XXII., fig. 6.

Geogr. distr.—Europe generally in the breeding season, ranging further northwards in the eastern than the western parts; in Asia as far eastward as E. Siberia; it winters in N. Africa: arrives in Great Britain in March, ranging along the southern coast as far as Cornwall, and up the eastern coast as far as Northumberland: in Scotland it is rarer.

Food.—Grass seeds, young shoots of aquatic plants, snails, small worms, larvæ, and aquatic insects.

Nest.—Bulky, carelessly constructed of flags, dried reeds, and leaves of aquatic plants; lined with fine grass.

Position of nest.—Upon a bed of broken reeds in marshy ground, and concealed amongst tall rushes.

Number of eggs.—9-12.

Time of nidification.—V-VII ; May.

Hewitson, on the authority of Mr. J. Hancock, states that "the eggs of the Spotted Crake, as well as those of the Water Rail, which are met with in exactly similar situations, are, in ordinary seasons, very difficult to obtain, the nest being placed in a thick bed of reeds, which covers a large extent of country, growing to a height of six or seven feet, and therefore not easily penetrated."—(Ill. Eggs Brit. Birds, vol. ii., p. 318.)

Owing to the drainage of much of the marshy land in the British Islands, the Spotted Crake has become much rarer and more local than formerly; nevertheless, wherever swamps and fens of any size exist it may be still found.

Both in appearance and habits this bird and the Water Rail are very similar, but frequently the eggs of the Spotted Crake more nearly resemble those of the Moorhen in colour, that which I have figured (selected by Mr. Dresser, as a typical form, from the series in his collection) being a case in point. At the same time the eggs figured by Hewitson, though much duller, and with darker markings, are more like those of the Water Rail, they were probably chosen as *striking* rather than *typical* forms from Mr. Hancock's series, since the egg is usually described as "ochreous, spotted and speckled with dark reddish-brown."

MOORHEN.

GALLINULA CHLOROPUS, *Linn.*

Pl. XXII., figs. 7-9; and Pl. XXIII., figs. 1, 2.

Geogr. distr.—Europe generally up to about 60° N. lat.: in Great Britain generally common and resident.

Food.—Grain, seeds of aquatic plants, tender grass shoots, aquatic insects, worms, and small Mollusca.

Nest.—Usually large and rather flat, formed of aquatic plants, reeds, and grasses, becoming somewhat finer towards the middle. I have, however, found very abnormal nests formed by turning over and plaiting together the leaves of tall reeds and flags, standing out of the water in a mill-pond, so as to form a rather deep cup, at the bottom of which the eggs were placed. I have myself taken eggs from a nest of this type.

Position of nest.—Usually amongst reeds near the edge of water, or upon broken down reeds in clumps standing on shoals in water; amongst aquatic herbage on banks, on mud-flats, amongst stumps and roots, or rarely in trees.

Number of eggs.—6-10 ; usually 8.

Time of nidification.—IV.

The Moorhen rears from two to three broods in the year. Mr. Hewitson remarks that its eggs, though very variable in size, "are subject to very little variation in colour." I think, however, that the figures on my plates, all from eggs in my collection, show enough variation to puzzle the young collector.

The nests, excepting the abnormal one above described, which is necessarily above the reach of water, and (for the same reason) such as are built in trees, are usually reeking wet, and take days of careful drying before fit for the cabinet. All that I have found were discovered in May, and none contained more than eight eggs ; but it is probable that the number may vary according to the brood. The variety (Pl. XXIII, fig 1) was taken by Mr. O. Janson from a nest at Braughing, Herts, and sent to me as a curiosity. I asked him to send me the remainder of the clutch ; these, however, turned out to be of the ordinary type figured at Pl. XXII, fig. 7.

When disturbed the Moorhen either takes wing or dives. After diving, especially when pursued by a dog, it remains submerged with only its beak above water, until all danger appears to be past.

COMMON COOT.

FULICA ATRA, *Linn.*

Pl. XXIII., fig. 3.

Geogr. distr.—Europe generally; Asia as far eastward as Japan; southwards to Central Africa; Australia generally distributed, common and resident in Great Britain.

Food.—Shoots of aquatic plants, &c., aquatic insects, and small Mollusca.

Nest.—A heavy and solid structure, though formed of loose but usually very wet materials, such as decayed aquatic herbage, decayed and fresh reeds, flags, &c., and well hollowed in the centre.

Position of nest.—Amongst reeds or willows, amongst grass or herbage on land, or, built to a good height, in shallow water.

Number of eggs.—5-8; rarely 12.

Time of nidification.—V-VI.

A beautiful nest of this species, obtained for me by the Hon. Walter De Rothschild, measures 15 inches by 10½ in diameter, and the central cavity 5 inches by 4.

The nest of the Coot is a common object on the Norfolk Broads, either in little islands of growing reeds or among the reeds which border their banks. As a rule the bulk of the nest is submerged, leaving only a few inches of the fresh reeds which form the upper layer above the surface.

In the autumn the Coots assemble in flocks, and frequently repair to salt marshes near the coast. They are, however, occasionally seen in winter upon ornamental water in or near cities, such, for instance, as those in Regent's Park and Kensington Gardens. I have myself seen it in the water in St. James's Park, where, however, it appeared to be semi-domesticated. Many birds, but especially the young chicks, fall victims to the rapacity of Pike, which drag them suddenly under water, whilst they are swimming, and devour them, whereas others are said to be swallowed by the Heron. It is said that the old birds sometimes repel the attacks of birds of prey by flapping water up at them with their wings.

FAMILY OTIDIDÆ.

GREAT BUSTARD.

OTIS TARDA.

Pl. XXXVI., fig. 2.

Geogr. distr.—Central and S. Europe, and eastward as far as Dauria : formerly tolerably common and resident in the British Isles; now almost extinct.

Food.—Seeds, corn, tender shoots, vegetable produce, pods of wild plants, insects, reptiles, &c.

Nest.—A mere depression in the soil, scantily lined with grass.

Position of nest.—Usually in a grain field: the eggs laid on the bare soil.

Number of eggs.—2-3; generally 2.

Time of nidification.—V; end of the month.

Although this species appears to have wholly ceased to breed with us since a time somewhere between 1830 and 1840, it has subsequently visited our islands from time to time; thus between the autumn of 1870 and the spring of 1871 more than a dozen Bustards, driven from the Continent by the unusually severe weather, are said to have visited Great Britain. It can, however, now be only regarded as an occasional straggler to our shores, and there is little hope that it will ever again remain to breed.

The question as to whether the great Bustard is polygamous or monogamous has yet to be definitely settled. Mr. Seebohm thus describes the finding of its eggs on the Wallachian Steppes :—" Although the Bustard is so wary, he often permits of the near approach of a waggon if the driver is concealed; and we soon had the pleasure of seeing a female Great Bustard rise from the grass, and, after a slight struggle, take wing and fly slowly away. We ran to the spot whence she rose, and were delighted to find two eggs on a piece of bare earth trodden down into the semblance of a hollow. There was no nest, and scarcely any cover; the grass was thin, and only here and there were weeds high enough to shield the sitting bird from view in certain directions. Whilst we were feasting our eyes on the eggs, she came round again, but, after one turn, flew right away, with slow, heavy flap of wing, not unlike a heron."—(Hist. Brit. Birds, vol. ii., p. 584.)

FAMILY ARDEIDÆ.

COMMON HERON.

ARDEA CINEREA, *Linn.*

Pl. XXII., fig. 1.

Geogr. distr.—Europe as far north as Central Scandinavia, Asia as far east as Japan, also Java; Africa as far south as the Cape; Australia: resident in Great Britain.

Food.—Worms, Mollusca, Crustacea, fish, reptiles, small birds and mammals, as rats and mice.

Nest.—Large, flat, roughly constructed of sticks lined with dry grass, wool, &c.; the centre of the nest has a cup-shaped depression for the reception of the eggs.

Position of nest.—In tall trees, upon cliffs of the sea-coast, or even occasionally on the ground.

Number of eggs.—3-6; rarely more than 4.

Time of nidification.—IV-V.

As is well known, these birds build in communities (called heronries), frequently represented by many nests, though I remember seeing a heronry in Kent represented by only two or three nests. In some parts of our country the increase of the Heron is checked on account of the havoc which the birds make amongst trout.

The old birds appear to rise with difficulty from a branch, though when soaring high their flight is powerful. Their action in the air always reminds me of that of a Gull, the body appearing to rise and fall between the wings, as though they were fixed at the tips.

Herons begin to repair to their nests early in February, leaving them again about the end of August, when they disperse through the marshy parts of the country. They usually rear two broods annually.

Touching the predacious habits of this bird Mr. Harting says :—" He will take young water fowl from the nest, and, after pinching them all over in his formidable bill, and holding them under water till they have become well saturated, he throws up his head, opens his mandibles, and the ' Moorhen souche ' disappears."—(' Sketches of Bird Life,' pp. 262-3.)

The egg which I have figured is printed rather too dark. It was given to me by the Rev. W. Bree, with other eggs, chiefly, I believe, collected by himself.

BITTERN.
Botaurus stellaris, Linn.
Pl. XXII., fig. 2.

Geogr. distr.—Throughout Europe; in Asia as far east as Japan; in Africa as far south as the Cape : formerly known to breed in England; but if it still does so it is but rarely.

Food.—Insects, reptiles, and fish.

Nest.—A tangled heap of dried reeds and flags, with rushes, grass, and sometimes a few sticks interlaced, the centre slightly depressed, and occasionally lined with the cotton of the reed.

Position of nest.—Generally amongst broken down reeds in the centre of fens and almost inaccessible morasses, sometimes upon diminutive islands rising out of dense growths of reeds and rushes.

Number of eggs.—3-4 ; rarely 5.

Time of nidification.—IV-V ; May.

It is possible that this species, which was at one time abundant enough in Great Britain, may still occasionally breed in some of the more extensive of our marshes. It is still found from time to time in various parts of the country, but probably the birds that are seen or shot are only chance stragglers. At the same time, as the Bittern is a particularly shy species, and but rarely leaves the dense cover of the reeds in which it dwells, it is possible that more birds may exist in the swampy districts of our islands than any one is aware of.

Hewitson says of its nesting habit :—" The Bittern makes its nest in the heart of fens and almost impenetrable marshy districts ; and, according to Dr. Thienemann, is careful to raise it beyond the effects of any temporary rising of the water, by placing it upon a mass of fallen reeds and prostrate rushes. The nest is formed of reeds, rushes, and grass, with occasionally a few sticks, slightly hollowed for the reception of the eggs, and sometimes lined with the cotton of the reed. The eggs are from three to five in number ; the time of incubation the month of May." —(Ill. Eggs Brit. Birds, vol. ii., p. 278.)

The egg which I have figured is in Mr. Dresser's collection.

LITTLE BITTERN.

ARDETTA MINUTA, Linn.

Pl. XXI., fig. 7.

Geogr. distr.—Temperate Europe generally in summer, and a rare straggler to Northern Europe; in autumn it migrates southwards, extending in winter tolerably far into Africa; Western Asia: in Great Britain it is a rare summer visitant, but has occurred in many parts of England, Scotland, and Ireland. It is said to have bred in England; specimens have been shot in Devonshire, one of which contained partly developed eggs.

Food.—Fish, Mollusca, and insects.

Nest.—A quantity of aquatic herbage, the foundation being mixed with twigs; lined with fine grasses and flags.

Position of nest.—Amongst reeds over shallow water, on the ground upon dry flags, or in a deserted Magpie's nest.

Number of eggs.—4-9; usually 5-6.

Time of nidification.—VI.

Though the breeding of this species in our islands has not hitherto been clearly proved, there is no reason that I know of why the occasional spring and summer visits of the bird to our shores, and more especially to the Norfolk marshes, should be entirely without result.

Mr. Seebohm says:—"The Little Bittern is very skulking in its habits, and frequents large marshes, swamps, clumps of bulrushes, and large expanses of reeds and rushes. Although it loves to frequent the solitudes of reeds it may sometimes be observed in the trees on the borders of the swamps, sitting quite still on the branches, with its neck stretched out and its beak pointing upwards."

"The nest of the Little Bittern is generally placed amongst the dense vegetation of its marshy haunts. Sometimes it is built amongst reeds a few inches above the water, and is often at a considerable distance from the shore. It is even said to sometimes take possession of a deserted Magpie's nest in a tree close to its haunt. The nest is very large for the size of the bird, loosely put together, and made of pieces of aquatic vegetation, sometimes a few twigs, and lined with finer material, such as grass or dead leaves of the reed."—(Hist. Brit. Birds, vol. ii., pp. 511, 512).

FAMILY **ANATIDÆ.**
MUTE SWAN.
CYGNUS OLOR, *Gmel.*

Pl. XXIII., fig. 4.

Geogr. distr.—Kept in a semi-domesticated state throughout Europe, but in some parts of N. and E. Europe tolerably numerous in a wild condition. Mr. Dresser says that it does not occur in a feral condition in Great Britain, the examples seen upon our coasts having strayed from swanneries. Generally distributed in England, but rarer towards the north; domesticated in Ireland.

Food.—Aquatic plants, Mollusca, fish and ova.

Nest.—A rounded mass of aquatic herbage or grass, four to five feet in width, heaped together; the central depression lined with somewhat finer materials.

Position of nest.—On a small island or large tussock rising out of the water, or close to the edge of a sloping bank amongst reeds.

Number of eggs.—5-8.

Time of nidification.—V.

Some twenty years ago, whilst on a visit to the Isle of Wight with a brother, we came suddenly upon a flock of thirteen Swans,* which were disporting themselves in and around a pool of shallow water not far from Alum Bay : at our approach they all took wing, with considerable noise, and, in my innocence, I concluded at the time that they were genuine wild swans; they were all in full adult plumage.

The Swan is very jealous of any approach to its nest, even though the eggs (owing to its own neglect) may contain dead cygnets; I remember well being chased by a pair of Swans, whose five eggs were in this condition, and I not only found it necessary to row hard, but also, from time to time, to defend myself with an oar from their vicious onslaught. They had not been on the nest previously for two or three days, and four of the eggs, when opened, were found to contain dead birds almost ready to hatch ; the fifth egg was addled, and was blown with ease.

---

* A note on the occurrence of a flock of fifteen at Newchurch occurs in the ' Zoologist,' vol. iii, p. 971.

GREY LAG GOOSE.

ANSER CINEREUS, *Meyer.*

Pl. XXV., fig. 1.

Geogr. distr.—Widely extended throughout the Palæarctic region ; wintering in N. Africa and India ; but not found in America. In Great Britain it breeds in the Outer Hebrides, and rarely in Ireland.
Food.—Seeds, grain, grass, and Mollusca.
Nest.—Loosely constructed of coarse dry grasses or flags, lined with feathers ; the eggs being also covered with down from the breast of the female.
Position of nest.—In a tuft of rank grass or heather.
Number of eggs.—4-7 ; usually 5.
Time of nidification.—IV-V.

The nest of the Grey Lag Goose is, when first made, tolerably well formed, but it gradually gets trampled down. The egg of this species was found by my brother, F. H. Butler, near Oxford, in May, 1873. The bird, which flew up at his approach, directed his attention to it. That this bird should breed so far south somewhat surprised me, but I do not think that I have hitherto recorded the fact. The bird was doubtless a mere straggler.

Although I have figured eggs of the two species of Geese which may breed with us, one illustration would answer for both. The egg of this species upon my plate is slightly abnormal in shape, being pointed at both ends : the specimen is in Mr. Dresser's collection. The two birds differ in the colour of their bills and legs, the bill of the Grey Lag being flesh-coloured, with the nail white, that of the Bean Goose orange, with the nail, edges, and base black ; the legs of the Grey Goose are flesh-coloured, but those of the Bean Goose orange.

Mr. Seebohm says that " a hundred years ago the Grey Lag Goose bred in the fens and marshes of the eastern counties of England ; but the reclamation of these extensive wastes has long since driven these birds to seek more congenial quarters. Its only breeding-places in the British Islands are in Scotland (in Ross, Sutherland, and Caithness) and on many of the western islands, and in Ireland on the lake at Castle Coole in Co. Monaghan."—(Hist. Brit. Birds, vol. iii.)

BEAN GOOSE.

ANSER SEGETUM, *Gmel.*

Pl. XXV., fig. 2.

Geogr. distr.—Europe and Asia, breeding in the north ; migrating southwards to winter in South Europe : according to Yarrell, it breeds in Westmoreland, Scotland, and the Hebrides, but this statement requires confirmation.

Food.—Seeds, grain, grass, aquatic plants, vegetables, Mollusca.
Nest.—Like that of the Grey Lag Goose.
Position of nest.—Same as that of the Gray Lag.
Number of eggs.—6-7.
Time of nidification.—VI.

Unless this species and its nest have been confounded with the Grey Lag, there seems no satisfactory reason for doubting that it does breed in Great Britain. At any rate the burden of disproving the existence of British eggs rests with those who disbelieve in them, a task which I fear they will find almost impossible, owing to the similarity of the nests and eggs, and the fact that the present species is stated to be very shy. At the same time, as **Mr.** Seebohm says, " there seems to be no evidence that it has ever bred in any part of the British Islands," although in spring and autumn it is a constant visitor to our shores.

Mr. Seebohm thus describes the nesting of this species :— " The Bean Goose repairs to the lakes on the Tundra, and chooses a hillock on the bank or an islet in the lake itself where the rushes and sedge are tall enough to conceal the sitting bird. A slight hollow is scraped in the soil, and lined with dead grass, moss, sometimes a few feathers, and always plenty of the light grey down of the bird itself. The number of eggs was generally three, but often four."— Hist. Brit. Birds, vol. iii., p. 495.)

SHOVELLER.

Spatula clypeata, *Linn.*

Pl. XXIV., fig. 1.

Geogr. distr.—Throughout Europe to Africa, Asia, and possibly Australia, and in America from Alaska to Costa Rica; common in Great Britain during the winter, a few pairs remaining to breed.

Food.—Worms, Mollusca, fish, Crustacea, insects, Algæ, grass.

Nest.—A mere hole scratched in the soil, lined with a little grass and a quantity of down from the parent birds.

Position of nest.—Close to water, especially fresh ponds and lakes, amongst the high grass, or under the shelter of a low bush.

Number of eggs.—9-12; occasionally 14.

Time of nidification.—V-VII; May.

Mr. Booth ('Rough Notes,' part IX.) says that "the rush-grown lochs in the east of Ross-shire are particularly attractive to this species, the character of the pools much resembling that of the Norfolk broads, where these birds are also resident. Further south than the swamps and flats of the eastern counties, I have not detected their breeding haunts, though Shovellers were annually seen during the winters I shot in Pevensey Level and Romney Marsh some twenty or five-and-twenty years ago.

"Shovellers seldom gather in large flocks; from a dozen up to twice that number may, however, occasionally be seen on the Norfolk broads."

Hewitson says that Mr. Hancock obtained two nests and eggs of this species upon Prestwick Carr, a piece of waste boggy ground near Newcastle, and that Mr. Salmon took it in Norfolk on the 10th of May; also (on the authority of the Messrs. Paget) that several nests, containing altogether fifty-six eggs, were found during one summer in Winterton Marshes in that county.

Though not abundant as a breeding species, at any rate where the district is not preserved, the Shoveller has been known to breed in Dorset, Kent, Norfolk, Hertford, Cambridge, Huntingdon, Yorkshire, East Lothian, Dumbarton, and Elgin, in Queen's County, Dublin, Donegal, and Antrim.

GARGANEY TEAL.

Querquedula circia, *Linn.*

Pl. XXIV., fig. 2.

Geogr. distr.—Greater part of Palæarctic Region ; India in winter, probably extending into the Indo-Malayan subregion; N. E. Africa: in Great Britain it is rare in Scotland and on the western coast of England, somewhat less so in the southern counties ; it breeds in Norfolk.

Food.—Insects, Mollusca, small fish, plants, and seeds.

Nest and position of same.—A mere depression in the ground in a meadow, or a grassy mound in a morass, usually near water, or, according to Hewitson, it is "composed of rushes and dried grass, mixed with the down of the bird, is placed upon the ground in low, boggy situations, among the coarse herbage and rushes in marshes, and on the borders of inland waters and rivers."

Number of eggs.—8-12, or even more.

Time of nidification.—IV-V ; May.

Mr. Seebohm says :—" The nest is placed in a variety of positions—hidden under a bush or in thick grass or sedge; far away from water in the forest or among the corn ; anywhere and everywhere where a hidden retreat can be found. At Riddagshausen, near Brunswick, I found a nest on the flat mossy margin of one of my friend Nehrkorn's lakes, without the slightest cover of any kind ; and Lord Walsingham showed me a nest near one of his lakes in South Norfolk in short heath. The nest is made very deep, and is lined with dead grass and leaves, to which is afterwards added plenty of down. The number of eggs varies from eight to twelve, or sometimes fourteen."—(Hist. Brit. Birds, vol. iii., p. 552.)

Mr. Harting says that " the Garganey, better known as the Summer Teal and Crick, or Cricket Teal, from its peculiar note, is an annual visitant to this country in spring, and in the eastern counties especially may be said to breed regularly." After speaking of its breeding in Norfolk, he adds, " In the adjoining county of Suffolk there is reason to believe that the Garganey nests every year in the marshes about Leiston, and the eggs have several times been found at Thorpe Mere, near Aldeburgh."

COMMON TEAL.

QUERQUEDULA CRECCA, Linn.

Pl. XXIV., figs. 5, 6.

Geogr. distr.—Palæarctic Region generally; everywhere in Great Britain.

Food.—Mollusca, fish, insects, seeds, plants.

Nest.—Constructed of heath and dried grass, thickly lined with down.

Position of nest.—On the ground amongst grass or heather, usually under the shelter of furze or other low bushes.

Number of eggs.—8-11.

Time of nidification.—IV-V ; May.

Respecting this species, Mr. Harting writes :—" I have often found the eggs during the last week in April. Although usually placed in the vicinity of water, the nest is sometimes at a considerable distance from it, and always rests upon dry ground. I have never found a Teal's nest in the swampy situations in which Coots, Moorhens, and Grebes build.

" A hollow is generally scraped out at the foot of some overhanging bunch of heather, or tussock of dry waving grass, and lined with fine heath stalks and bents, then eight or ten creamy white eggs are laid, and as the hen bird covers them she plucks from her breast and sides the soft brown down which underlies her feathers, and places it entirely round the eggs, filling up all the interstices, thus forming a warm bed for the young as soon as they leave the shell. The overhanging roof of grass or heather serves to conceal the eggs (though not always so) from the eyes of passing Crows."

In his ' Sketches of Bird Life,' p. 279, he says :—" Teal are very sociable in their habits, and during the winter they may often be found in company with the common Wild Ducks. But, although they mingle together when on the water, on being disturbed the species always separate, the Teal going off in one flock, the Ducks in another. On rising from the water they do not first swim away from the danger, or flutter over the surface as some fowl do, but jump suddenly into the air without warning, so that if you are approaching them in a punt and do not pull the trigger the moment you find that you are perceived by them, you will, in all probability, lose your chance of a shot."

COMMON SCOTER.

ŒDEMIA NIGRA, Linn.

Geogr. distr.—Throughout the northern portions of Europe and Asia, ranging southwards during the winter, when it is widely distributed in Great Britain : it is said to breed in several parts of Scotland, such as the moors in Caithness and Inverness.*

Food.—Algæ, insects, Crustacea, Mollusca, worms.

Nest.—A mere depression or hollow scratched in the ground, and lined with moss, grass, and down.

Position of nest.—Usually on small islands in open swampy ground or under the shelter of a bush.

Number of eggs.—8-9.

Time of nidification.—VI ; middle of the month.

Mr. Cecil Smith, in his 'Birds of Somersetshire,' states that the Scoter is a very expert diver, and will remain under water for a long time. These birds are not known to breed in England, but repair to more northern latitudes for that purpose. (See pp. 501, 502.)

Mr. Harting says that this species has but little claim to be protected by the Wild Fowl Preservation Act, "although there is some little evidence on record of its occasionally breeding in the British Islands. Mr. A. G. More, writing of this species in the 'Ibis' for 1865 (p. 445), says :—'Mr. W. Dunbar tells me that the Black Scoter breeds every year in many parts of the moors of Caithness, making its nest in the boggy swamps around the lakes. He has known the eggs taken more than once. Mr. R. J. Shearer writes that a "black duck" is well known as breeding on one or two lakes in the Thurso district.' In reply to my suggestion that the bird referred to by Mr. Shearer might be the Tufted Duck, Mr. More wrote me that, since penning his former remarks, the nest, with young and eggs, had also been taken, and old birds killed at the same time. This, of course, places the identity of the species beyond doubt. Mr. More added that in 1871 Small, the bird-stuffer in Edinburgh, showed him some very young Scoters which had been obtained, with the old birds, in Caithness."

---

* I quote Dresser's 'Birds of Europe' as authority; Hewitson (in his 'Illustrations') was evidently unaware of the fact : the egg, as figured by him, is similar to my figure of the Garganey.

WILD DUCK (OR MALLARD).
ANAS BOSCHAS, *Linn.*
Pl. XXIV., fig. 3.

Geogr. distr.—Europe generally, from Northern Asia to China and Japan; Northern Africa; N. America as far southwards as Mexico: in Great Britain more commonly in the north than the south during the summer; resident in Ireland.

Food.—Plants, seeds, grain, insects, worms, Mollusca, small fish and ova.

Nest.—A roughly-made and bulky structure formed of grasses, sedges, flags, and other plants.

Position of nest.—Rarely far from water; either amongst reeds on the edge of a marsh or in heathy meadows, frequently under a bush; in woods; also on stumps of pollard willows, or in deserted nests of various species of *Corvidæ.*

Number of eggs.—7-12.

Time of nidification.—IV-V.

Captain Feilden says that this species is common as a breeding species throughout the Faroe Islands, every considerable piece of fresh water being tenanted by a pair or more.—('Zoologist,' 1872.)

Mr. W. J. Sterland, in his 'Birds of Sherwood Forest,' speaks of this bird as nesting at some height from the ground, and states that the young are carried to the ground by the female. He says:—"The late Mr. Mansell, of Thoresby, related to me the following instance of the parent bird actually conveying her young to water, which he himself witnessed. He was passing one morning at daybreak, in the early part of May, under a large ash tree, which was thickly clothed with ivy, when a cheeping and rustling overhead induced him to withdraw a few steps and stand still. He had hardly done so when a Wild Duck flew out of the ivy, some height up the tree, holding a young one in her bill. This she put down on the bank of the stream, which was about a hundred yards from the tree, and then, returning to the nest, conveyed the remainder, one by one, in the same manner, until thirteen were safely placed on the bank. Here she brooded them for a few minutes, and then, with much apparent fondness, led them down the bank into the water, where they were speedily darting about with the utmost liveliness."—(Pp. 219, 220.)

COMMON SHELDRAKE.

TADORNA CORNUTA, *Gmel.*

Pl. XXIV., fig. 4.

Geogr. distr.—Europe generally, eastward across the continent of Asia to China and Japan, southward to N. Africa ; resident in Great Britain from the southern counties to N. Scotland, but scarce in the south during the breeding season.

Food.—Algæ, insects, Mollusca, fish and ova.

Nest.—A little dried grass with a lining of down, or entirely of down.

Position of nest.—In deserted rabbit-burrows, or sometimes in the hole of a Badger or Fox, at a distance of from three to six feet from the entrance, which is covered with a sod made like a lid.

Number of eggs.—7-16 ; rarely more than 12.

Time of nidification.—V-VI.

A handsome species frequenting the sea-shores ; breeding in Norfolk among the sand-hills on the coast. According to Mr. Cecil Smith, the birds collect in considerable numbers at their various breeding stations, from about May till July or August, after which they become scarcer in the immediate locality which they had selected. He states that, although the species usually breeds in a rabbit's burrow, nevertheless, " if a convenient hole cannot be found, the nest is occasionally placed in a thick bramble, or furze-bush,—always in the very thickest part,—a regular creep being made, through which the bird approaches her nest, and which the eager birds-nester will have to follow up for some distance before he will be able to reach the eggs."—(' Birds of Somersetshire,' pp. 676, 677.)

In the 4th edition of Yarrell, this bird is said still to breed " sparingly in the rabbit-burrows and sand-hills upon the coast of Suffolk, Norfolk, Lincolnshire, Yorkshire, Durham, and Northumberland ; also in suitable localities along the east coast of Scotland, where large flocks are observed in winter along the sandy estuaries. To Shetland it is a rare visitor at any season ; but is more common in Orkney. It is numerous in summer in the Hebrides, and also in some districts on the west side of the mainland. The increase of population has, of course, acted unfavourably to it in the north-west of England, but it still breeds along the sandy coasts of Lancashire, Cheshire, and many parts of Wales. In

Cornwall, according to Rodd, it is only a winter visitant; but a few nest in Devonshire, Somersetshire, and near Poole Harbour in Dorsetshire. In Ireland it still nests in suitable localities, among which may be mentioned the sand-hills of Bartragh, Co. Mayo (owing to strict preservation by the owner, Captain Kirkwood), the Saltees, portions of the Wexford coast, Dungarvan and Tramore Bays, Co. Waterford."—(Vol. iv. p. 353.)

WIGEON.
MARECA PENELOPE, *Linn.*

Pl. XXV., fig. 3.

Geogr. distr.—Throughout Europe, breeding in the northern districts and wintering in the southern ; Asia as far eastward as Japan;
has been found on the eastern coast of America : a common winter
visitant in Great Britain, but rarely breeds with us, so far as is known;
it however probably breeds in the Outer Hebrides, &c., and in Norfolk.

Food.—Seeds, grass, water-plants, and Mollusca.

Nest.—A mere depression or hole scratched in the ground, lined
with down, a few feathers, decayed reeds or moss, and grass-bents
matted together.

Position of nest.—Sometimes close to water, but by no means
invariably; placed amongst rank herbage, and low bushes or long
rushes.

Number of eggs.—5-12 ; usually 5-8.

Time of nidification.—V-VI.

Usually breeds at the end of May or the beginning of
June. Mr. Cecil Smith says that "the nest of a Wigeon
which was found by Mr. Selby, on one of the islands in
a lake in Scotland, was placed in the heart of a large
rose-bush, and was made of decayed rushes and reeds,
with a lining of warm down from the bird's body.

" The food of the present species consists almost entirely
of grass, which, to judge by tame ones, it eats most
greedily, as these spend nearly the whole of the day in
cropping the short grass on the lawn. Meyer adds—
' aquatic insects and larvæ, worms and small Mollusca,
rarely the small fry of fish and frogs:' he also says—
' it occasionally eats grain, but that it is no favourite
food of the Wigeons,' which seems certainly to be the case.
I have known my tame ones reduced to eat grain with
the other wild fowl during deep snow, when they could
get no grass, but at other times they do not appear to care
at all about it."—('Birds of Somersetshire,' p. 498.)

Although it is probable that this species has bred in
Norfolk, there is at present no proof of its having done so
either in that or any other county of England or Wales ;
but in Ireland, Scotland, and the Scotch islands it is known
to breed.

GOLDEN EYE.
CLANGULA GLAUCION, Linn.
Pl. XXVI., fig. 1.

Geogr. distr.—Northern Palæarctic and Nearctic Regions, migrating southwards as far as the northern shores of Africa to winter ; may breed in the Shetland Isles. According to Gray, it probably breeds occasionally in Sutherlandshire.
Food.—Algæ, Mollusca, Crustacea, and fish.
Nest.—A hole or depression lined with down.
Position of nest.—Either in a hollow tree at some height from the ground, under brushwood, or amongst rushes at the edge of water.
Number of eggs.—10-19.
Time of nidification.—V.

The Golden Eye is a constant winter visitant in Great Britain. Its favourite nesting place is a hole in a tree growing near water. Mr. Booth, in Part VIII. of his ' Rough Notes,' says, respecting this bird :—" Though much time was spent in the attempt, I never succeeded in verifying the fact of this species breeding in Great Britain. The female I repeatedly observed during summer on remote Highland lochs, and on more than one occasion a bird was detected flying from old and weather-beaten pine-woods, where, doubtless, her nest was concealed ; sweeping rapidly beneath the branches in the shade thrown by the dense timber, it was by no means easy to keep the small grey-tinted fowl in view. For several days subsequent to the 11th June, 1869, I observed a male, at times in company with a female, disporting himself in the water at the east side of Loch Slyn in Ross-shire. While watching the pair the female on one occasion disappeared without attracting attention ; shortly after, however, she came in sight, skimming from the plantation of Scotch firs standing to the east of the loch, and rejoined the drake, when both birds, evidently disturbed by the approach of the punt, rose on wing and left the water. The male evinced a decided aversion to permit of a close inspection ; judging, however, from the nearest view I was enabled to obtain, he appeared to be in full adult plumage. This is the only instance when the mature drake came under my notice later than the first week of April."

POCHARD.

FULIGULA FERINA, *Linn.*

Pl. XXVI., fig. 2.

Geogr. distr.—Europe generally, extending southwards to N. Africa and eastward as far as China; not rare in Great Britain; breeds in Norfolk, Yorkshire, Dorsetshire, &c.

Food.—Mollusca, Crustacea, fish, aquatic plants.

Nest.—A mere depression in the ground, lined with dried vegetable matter and down or feathers.

Position of nest.—In marshy or reedy places near water.

Number of eggs, 7-8.

Time of nidification.—V-VI.

Mr. Booth says that " in the eastern counties Pochards exhibit a decided partiality to the society of Coots; it is seldom that the two species are seen on the same broad unless in company. Some twelve or fifteen years ago, I often watched immense bodies feeding in company, two or three acres of water having been at times almost black with birds; their number, however, unless alarmed and put in motion, could scarcely be estimated, as many were constantly disappearing below the surface."

" Though I failed to detect the nest," he continues, " and never met with a chance of examining the young in down, it is well known that Pochards breed in more than one locality in Great Britain. A few stragglers continue in the Highlands during April, but after that date the species is seldom observed. On two or three occasions, however, between the 11th and 20th of June, 1869, when visiting Loch Slyn, I noticed a couple of drakes on the water. These birds appeared unusually regardless of danger; and, anxious to know whether they remained of their own free will, or their presence at this season was the result of wounds, I pressed them closely in the punt, when rising at once on wing, they made the circuit of the loch, and after a short flight returned to the water. Judging by their actions, I am of opinion that the females were sitting on some of the marshy spots near at hand."—(' Rough Notes,' Part VIII.)

Mr. Cecil Smith says that " the nest is placed amongst reeds, rushes, and other coarse herbage, on the borders of inland waters. Meyer says that in localities that suit the species great numbers of nests are placed near each other, although each nest is placed out of sight of the others."—(' Birds of Somersetshire,' p. 504.)

TUFTED DUCK.

FULIGULA CRISTATA, *Leach.*

Pl. XXVI., fig. 3.

Geogr. distr.—Europe generally, but breeding only in the north; migrating into N. Africa to winter; Asia as far eastward as Japan, and southwards into S. India. In Great Britain it is principally a winter visitant; but it has been recorded as breeding in Sussex, Surrey, Shropshire, Nottinghamshire, Yorkshire, and Perthshire; also in several localities in Ireland.

Food.—Aquatic insects, Mollusca, fish, tadpoles, frogs, and Algæ.

Nest.—Formed of grass, bents, and a few leaves felted together with a mass of down.

Position of nest.—On the ground not far from fresh water, under a bush or tuft of sedge.

Number of eggs.—8-13.

Time of nidification.—V-VI.

Mr. Cecil Smith says of this species :—"It is easily kept in confinement, and is very conspicuous and ornamental. It has also been known to breed in confinement in the Zoological Gardens, though in a wild state it seems very rarely, if ever, to breed in any part of Britain. In a neighbouring pond a wild female Tufted Duck remained for a long time into the spring in company with two pinioned males, but did not remain to breed.

"Meyer says 'the nest is usually placed in a hollow on grassy ground, or under shelter of a stone, or stump of a tree, or of some vegetable production, at a distance of from sixty to a hundred yards from the water : it is made of stalks and grasses carelessly put together.'

"The food of the Tufted Duck, which is mostly obtained by diving, consists of shell-fish, small frogs and their spawn, aquatic insects, and the roots, buds and seeds of aquatic plants. In confinement it will also eat grain, and both this bird and the Pochard may be easily got to come to the call and dive for grain or bread."—('Birds of Somersetshire,' pp. 508-9.)

The call-note of this species is said to resemble the words "*currugh, currugh*" uttered gutturally.

EIDER DUCK.
SOMATERIA MOLLISSIMA, *Linn.*

Pl. XXVI., fig. 4.

Geogr. distr.—Common in the Western Palæarctic Region, less so in the Eastern portion. In Great Britain it occurs in the Hebrides, Farn, Orkney, and Shetland Islands, also on the coasts of Scotland and Northumberland,

Food.—Seaweed, Mollusca, Crustacea, and fish.

Nest.—A mere depression in the ground, lined with fine seaweed or dried grass, down, heather, or small twigs.

Position of nest.—Usually amongst heather or nettles, under a large stone or juniper bush.

Number of eggs.—5-10, or even more.

Time of nidification.—VI.

Mr. Harting says that Eider Ducks "are very common round Mull and Iona, and on the shores of North and South Uist, Benbecula, and Lewis, and numbers breed upon the islands in the Sound of Harris. The southernmost breeding station is believed to be Coquet Island."

Mr. Howard Saunders says :—"Although the nest of the Eider is usually at no great distance from the water, it has occasionally been found a mile or two inland, and also at a considerable elevation."—(Yarrell's Hist. Brit. Birds, 4th ed., vol. iv., p. 459.)

Major Feilden (' Zoologist,' 1872, p. 3254) mentions having met with a nest of this duck in the Shetlands which was placed in the midst of knee-deep heather, at least five hundred feet above the sea level.

This duck always covers up its eggs with grass and leaves before absenting itself from the nest; this is partly done, no doubt, to retain warmth in the eggs, but the principal object is obviously to conceal them from view.

GOOSANDER.

MERGUS MERGANSER, Linn.

Pl. XXVII., fig. 1.

Geogr. distr.—Entire Palæarctic and Nearctic Regions, being found in the north in summer and in the south in winter ; to Great Britain it is for the most part a winter visitor, though it occasionally breeds in Perthshire, and possibly the Outer Hebrides.*

Food.—Fish, Amphibians, and Crustacea.

Nest.—A hollow scratched in the ground, or more often in a hollow tree, lined with dried grass or rootlets, down, and sometimes husks of beech-buds.

Position of nest.—Near large lakes, or on small islands in the vicinity of the sea.

Number of eggs.—8-12.

Time of nidification.—IV-V ; usually the end of April or beginning of May.

In Part VII. of his 'Rough Notes' Mr. E. T. Booth says:— " Throughout the districts in which I met with Goosanders during the breeding season, the females appeared in some instances to resort to situations for nesting purposes at a considerable elevation on the hills. A cavity in a large and partially decayed birch was pointed out by a keeper as a spot from which some eggs (previously seen in his possession) had been taken. The old and weather-beaten stump was on the outskirts of a thicket of birch, fir, and alder stretching from a swamp up a steep brae, and within a mile of a loch in which I have repeatedly watched two or three broods. The tree was carefully examined, and I noticed that down from the breast of the bird was still clinging to the rotten wood ; the general appearance also of the rubbish in the hollow left little doubt as to the truth of the statement. On more than one occasion I have been informed by keepers and gillies well acquainted with this species that they had met with broods on the bare and open moors following the course of some of the larger burns."

* Mr. Howard Saunders, however, records his conviction that the eggs taken in the Hebrides were those of the Red-breasted Merganser.

RED-BREASTED MERGANSER.
MERGUS SERRATOR, *Linn.*
Pl. XXVII., fig. 4.

Geogr. distr.—North Europe, Asia, and America in summer; North Africa, China and Southern N. America in winter: in Great Britain it occurs chiefly in autumn and winter, though it breeds not infrequently in certain parts of Scotland, and also in Ireland.

Food.—Fish, Amphibians, and Crustacea.

Nest.—Formed of fine grasses, twigs, and moss, matted together and mixed with down.

Position of nest.—On the ground in quiet nooks, amongst bushes or rank vegetation, or in the hollow of a tree.

Number of eggs.—8-12.

Time of nidification.—VI.

In Mr. Booth's 'Rough Notes,' we read that "These birds for the most part resort to the open sea, though when undisturbed they not infrequently make their way to the small pools of brackish water inside the sea-beach, and occasionally proceed some distance up the rivers. The nest is usually carefully concealed beneath an overhanging slab of rock, or in some slight cavity in a heathery bank ; at times, on the shores of inland lochs, I have seen them well hidden among thick bushes and shrubs, or under the stem of some rugged and weather-beaten tree. The eggs, most frequently from eight to ten in number, are warmly covered with the grey down from the breast of the female. The outer portion of the nest is composed of dried strands of coarse grass, with occasionally a few small heather stalks." —(Part V.)

Mr. Harting says that this species " is a native of nearly all the lakes of any importance north of Loch Lomond ; it is likewise numerously distributed throughout the Long Island, where it appears to be permanently resident, breeding in South Uist, North Uist, Benbecula, the islands in the Sound of Harris and Lewis. Within the circle of the inner islands it is found breeding on rocky islets off Skye, Islay, Jura, Colonsay, and Tyree. In Orkney the Merganser continues the whole year, and may be seen every day in numbers on the lakes and in the sea." Mr. Harting adds that it breeds in Shetland and on various islands and lakes of Ireland."

Family PODICIPITIDÆ.

GREAT CRESTED GREBE.

Podiceps cristatus, Linn.

Pl. XXVII., fig. 2.

Geogr. distr.—Temperate portions of Continental Europe, Asia as far eastward as Japan, Africa southwards to the Cape, but rare in Egypt, Australia, and New Zealand; it is less common in Great Britain than formerly, but still breeds in Hertfordshire, Suffolk, Norfolk, Huntingdonshire, Warwick, Worcestershire, Shropshire, Lincolnshire, Cheshire, Yorkshire, and Wales, also rarely in Scotland and Ireland.

Food.—Aquatic plants, worms, insects, Mollusca, fish and ova, Amphibia.

Nest.—A mass of aquatic herbage, reeds, grass, &c., of a circular form, and about ten inches in depth, with slight central depression.

Position of nest.—Floating deeply in the water, only three or four inches being visible in those which I have seen, anchored amongst growing reeds and flags.

Number of eggs.—4; rarely 5.

Time of nidification.—IV-VII.

In the last week of May, 1885, when on Norfolk Broads with Mr. O. Janson, we came across several nests of this species. As we were too early for the second brood, only one nest was taken with two eggs.* When removed from the water it was as much as a man could lift. All the nests which we saw were formed of masses of leaves of reeds and flags, mixed with feathery green weeds, and bound at the borders with fresh green reeds; from the middle of the central depression a number of long, thin, twisted grasses trailed over the upper surface, giving the whole nest a lively appearance,† whilst the eggs were concealed by pieces of the leaf of the water-lily. The birds in Norfolk are known as "Loons."

In Africa this Grebe is said to breed in companies consisting of six to eight nests, whereas in England they appear to fight shy of one another.

---

* In the first week of June of the present year we took the third of three nests (ten eggs in all); doubtless there were double this number on the same Broad.

† Later in the year the nests exhibit a miserably decayed appearance, but the birds do not seem to object to use them in this state.

LITTLE GREBE (OR DABCHICK).

PODICEPS FLUVIATILIS, *Tunst.*

Pl. XXVII., fig. 3.

Geogr. distr.—Europe generally from Scandinavia to Africa, and on the latter continent to the Cape; Asia to Japan, and southwards to Australia and New Zealand; generally distributed and resident in Great Britain.

Food.—Water-plants, seeds, insects, Mollusca, fish.

Nest.—A mass of aquatic herbage, such as reeds, flags, rushes, mosses, and soft, green water-weeds, sometimes piled up to about a foot above the surface of the water.

Position of nest.—Usually amongst reeds or rushes, sometimes floating but anchored to the growing stalks; in lakes, slow rivers, and occasionally fish-ponds.

Number of eggs.—4-5; rarely 6.

Time of nidification.—IV-VII; May.

Although the top of the nest is, as already stated, occasionally raised some height from the water, I believe it is rarely more than an inch or two above it. It is always saturated with moisture, like that of *P. cristatus*, and is covered with freshly-gathered weeds. Its position is generally from twenty to thirty yards from the water's edge. When disturbed, the bird almost invariably dives; it is rare indeed for it to attempt to escape by flight. If forced to leave its eggs, it invariably, before diving, covers them up with fresh weed from the side of the nest.

When first laid the eggs of the Little Grebe are pure white, but they gradually lose their clean aspect, probably being purposely stained by the bird with wet and partly-decayed weed. I have seen an egg which was taken from the nest as soon as deposited, the bird having been surprised in the act of laying, and it was as pure in colour as that of a Wood Pigeon.

According to Mr. Charles Thurnall (see 4th ed. Yarrell, vol. iv., p. 139) this bird, when accompanied by young chicks, and pursued, appears to tow them along—" one went on one side and one on the other, and seemed to take hold of the feathers beneath the wing, which she raised apparently for that purpose, and placing the tip of a wing on each young one, as if to keep them in their places, she swam slowly off."

FAMILY **COLYMBIDÆ.**

RED-THROATED DIVER.

COLYMBUS SEPTENTRIONALIS, *Linn.*

Pl. XXVIII., fig. 1.

Geogr. distr.—Northern portions of Europe, Asia, and America in summer, ranging southwards to N. Africa, China, and Maryland to winter; in Great Britain it is resident and not uncommon, breeding in the north and wintering in the south.

Food.—Frogs, fish, insects, and aquatic plants.

Nest.—A small depression in the ground, lined with a few rushes and a little grass.

Position of nest.—Near the water's edge, usually on small islands or the margins of inland lakes.

Number of eggs.—2.

Time of nidification.—VI; early in the month.

According to a statement in one of the first volumes of the ' Zoologist,' this species, upon its arrival on our shores, flies to and fro generally in small flocks, which form two lines about a mile apart, and the nearer line about a mile and a half from the shore : distances are, however, very difficult to measure with the eye, as will at once be discovered by any Englishman visiting Switzerland for the first time. I well remember when at Chamounix seeing a hawk pounce upon a small bird, at apparently a distance of about half a mile, but which my guide assured me was at least five miles; the truth probably lay somewhere between the two measurements.

Mr. J. A. Harvie Brown says that in Scotland, where both this and the Black-throated species are by no means uncommon on fresh-water lochs, the Red-throated Diver very rarely, if ever, breeds upon the larger lochs, preferring the quiet moorland, marshy-edged tarns, and " lochans," often nesting on the shore, and only frequenting the larger lochs in search of food. It is rare for two pairs to be found breeding upon the same loch.—(See 'Zoologist,' 1877, p. 296).

Mr. Cecil Smith says that this species sometimes places its eggs so near the water as to be able to reach it with its bill.—(See ' Birds of Somersetshire,' p. 544.)

BLACK-THROATED DIVER.

COLYMBUS ARCTICUS, Linn.

Pl. XXVIII., fig. 3.

Geogr. distr.—Northern portions of the old and new continents generally in summer, ranging southwards in winter; occurs sparingly on the English coast in cold weather; in Scotland, where it breeds, it is resident.

Food.—Fish, Mollusca, Crustacea, insects, aquatic plants.

Nest.—Generally none, the eggs being placed in a mere depression on the grass or moss; but sometimes this is scantily lined with weeds or sedge.

Position of nest.—On a small islet or spit of land projecting into a lake or inland loch; near the water's edge.

Number of eggs.—1-2.

Time of nidification.—V-VI.

In choosing a nesting-place, this species, according to Mr. E. T. Booth, frequently selects a small loch in preference to a large one, but sometimes the latter is selected, particularly if dotted with grass-covered islets having shelving banks.

According to Mr. Howard Saunders, "The Red-throated Diver frequents more retired spots than the Black-throated species, and prefers the shores of small tarns, or even pools—often at a considerable elevation—to the islands of a larger loch. Sometimes there is a slight nest of loose rushes and dried grass, but more frequently the eggs are laid upon the bare turf, or stones, within a few feet of the water's edge."—(Yarrell's Hist. Brit. Birds, 4th ed., vol. iv., p. 114.)

Mr. Seebohm, however, merely says that "The Black-throated Diver breeds somewhat sparingly and locally in the lochs of the Outer Hebrides, and in the counties of Argyll, Perth, Inverness, Ross, and Sutherland."—(Hist. Brit. Birds, vol. iv., p. 407.)

GREAT NORTHERN DIVER.

COLYMBUS GLACIALIS, *Linn.*

Pl. XXVIII., fig. 2.

Geogr. distr.—Almost circumpolar: suitable localities throughout N. Europe, Asia, and America, excepting N.W. America; in winter extending in the Palæarctic region as far south as the Mediterranean, and in America as far down as Texas; in Great Britain it occurs sparingly in autumn and winter, being not uncommon in the South of England. It is said to breed in Scotland, and Dr. Saxby believes that it breeds in the Orkneys.

Food.—Small fish and Crustacea.

Nest.—A heaped-up mass of aquatic vegetation and grass.

Position of nest.—On small islands or collections of half-floating aquatic plants near edge of water.

Number of eggs.—2-3; usually 2.

Time of nidification.—VI.

According to Mr. Cecil Smith, the Great Northern Diver, though not known to him to breed in England, probably does so occasionally in the vicinage of inland waters in the extreme north of Scotland and in the Scotch Islands.

"The nest, which is flat, and made up of dead herbage, is placed near the water, amongst reeds and flags. The nesting time appears to be the only period in which this bird ever willingly goes on shore, for, as may be at once seen from the position of its legs and feet, it is not a great pedestrian, and, although a good flyer when once on the wing, it cannot rise from the land: when on that element it seems to progress more like a seal than a bird, jumping along on its hind legs (!) and wings."—('Birds of Somersetshire,' pp. 537-8.)

The eggs of this and the Black-throated species were lent to me for illustration by Mr. Dresser; that of the Red-throated Diver was given to me by the Hon. Walter de Rothschild.

FAMILY **ALCIDÆ**.
PUFFIN.
FRATERCULA ARCTICA, *Linn.*

Pl. XXIX., fig. 1.

Geogr. distr.—Northern portions of Europe, breeding as far north as the coasts of France, and in winter straggling down to the shores of N. Africa; Atlantic coasts of N. America. In Great Britain it is not uncommon, though more frequent in the north than the south.

Food.—Fish, Mollusca, Crustacea, and Algæ.

Nest.—If any, consisting only of a small quantity of grass, in a hole two or three feet deep.

Position of nest.—On tolerably high turf-covered table-land, on cliffs overhanging the sea, in crevices of rocks, or in deserted rabbit-burrows.

Number of eggs.—1.

Time of nidification.—V-VI.

Respecting the distribution of the Puffin in the United Kingdom, Mr. Seebohm remarks :—" The Puffin is one of the best known of British sea-birds, and is found in all suitable localities along the entire coast-line of our islands during summer. In rocky districts it is much more numerous than on low-lying coasts, and it is especially abundant at Lundy Island, Priestholm, off the coast of Anglesey, the Isle of Man, the Hebrides (especially St. Kilda), the Orkneys, Shetland, and the Farne Islands. It is also equally common on the Irish coasts."—(Hist. Brit. Birds, vol. iv., p. 364.)

Speaking of the breeding of this species in an island in Norway, Hewitson says :—" The island, which sloped gradually upward from the edge of the water to the base of a lofty cliff, was entirely covered with large fragments of rock piled upon each other in the wildest confusion, and under them vast numbers of these birds were breeding. Many thousands were passing in rapid flight around us, and thousands more were underneath our feet ; as we stumbled onwards we could distinctly hear them as we passed over their heads, croaking and gabbling to each other, and no doubt complaining of our intrusion."—(Ill. Eggs Brit. Birds, vol. ii., pp. 408-9.)

RAZORBILL.

ALCA TORDA, Linn.

Pl. XXIX., figs. 2-3.

Geogr. distr.—Northern Europe, the coasts of Japan, the North Pacific Ocean, and the northern parts of eastern N. America; rarely in Algeria. In Great Britain it is resident from the extreme north of Scotland to the south of England, breeding in great numbers at Flamborough Head, the Orkneys, Shetlands and Hebrides.

Food.—Fish and Crustacea.

Nest.—None.

Position of breeding-place.—On the bare unsheltered surface, or in a hole or cranny on ledges of high, rugged rocks overhanging the sea.

Number of eggs.—1.

Time of nidification.—V-VI; May.

Mr. Howard Saunders observes that, " as a rule, the Guillemots occupy one station or line of ledges on the rock; the Razorbills another; the Puffins a third; the Kittiwake Gulls a fourth; whilst the most inaccessible crags seem to be left for the use of the Herring Gulls. The Razorbills generally select the higher and rougher ledges, and they are partial to crevices, their eggs being sometimes deposited so far in that it is no easy matter to get at them; at other times they lay their eggs on the broader shelves along with the Guillemots, but not so closely together. At the Farne Islands (he) once saw a Razorbill sitting on her egg in the old nest of a Cormorant. When incubating they lie on their eggs, the mate often standing by the side of the sitting bird. Sometimes the male brings food to the female, but both birds take their turn at incubation, one having been seen to fly to the sitting one and give it a gentle peck, when the latter immediately ceded its place."—(Yarrell's Hist. Brit. Birds, 4th ed., vol. iv., p. 56.)

Mr. Theodore Walker (Zool., 1871, p. 2427) says that he has seen the Razorbill, seizing its young by the back of the neck, convey them to the sea, where, by repeatedly carrying them under water, it teaches them to dive.

COMMON GUILLEMOT.
LOMVIA TROILE, *Linn.*
Pl. XXX.

Geogr. distr.—Northern portions of Europe and the Atlantic shores of N. America, extending further southwards in winter; common and resident on all the British coasts, breeding abundantly on the Fern Islands, St. Kilda, and other suitable localities.

Food.—Fish, Mollusca, and Crustacea.

Nest.—None.

Position of breeding-place.—On narrow ledges of bare rock, usually high above the water.

Number of eggs.—1.

Time of nidification.—V·VI.

This bird, according to Mr. Cecil Smith, " is very common, and resident throughout nearly the whole of the English coast, collecting at the various breeding stations in the summer, and spreading about in various parts of the sea in search of food as soon as the young are sufficiently advanced to take care of themselves. At the breeding stations they may be seen in thousands, some on the water engaged in fishing, and others standing in long rows on the rocks, lining every available ridge like detachments of skirmishers.

" The eggs of the Guillemot are placed on the bare ledges of the rocks, without any nest, and often in very exposed situations, where it seems wonderful they are not blown off by the wind, but probably their pear-shape protects them from this, as they only run round when moved by the wind.

" The parents are said to carry their young down to the water on their backs, but I have never seen this done, though I have seen young birds in the water that certainly could not have flown down, and a fall from the height at which the eggs are usually placed would have been fatal to the young birds, even if they fell into the water."—(' Birds of Somersetshire,' pp. 547-9.)

BLACK GUILLEMOT.
URIA GRYLLE, *Linn.*
Pl. XXXI., fig. 2.

Geogr. distr.—Northern portions of the Palæarctic and Nearctic regions, migrating southwards in winter, but not very far. In Great Britain it is rare on the English coast during the nesting season, a few laying on the Welsh coast near Tenby, and in the northern counties; but in Scotland and Ireland it is tolerably common, breeding on the western coast of Scotland, in the Orkneys, Shetlands, Hebrides, and in the Bay of Dublin, and also sparingly in the Isle of Man.

Food.—Young fish, Mollusca, and Crustacea.

Nest.—None.

Position of breeding-place.—In hollows or crevices of the rocks at considerable altitudes, or amongst rocks and stones on beaches, or beneath large stones or boulders upon grassy islands.

Number of eggs.—2-3; usually 2.

Time of nidification.—V.-VI.

The following account of the nesting of this species I take from Mr. Seebohm's work, vol. iii., p. 385 :—

" The Black Guillemot probably pairs for life, as year by year the same crannies and crevices of the rocks are tenanted, presumably by the same birds. It is a rather late breeder, its eggs being seldom deposited in Scotland before the end of May or the first week in June, and fresh eggs may be obtained all through the latter month. It makes no nest, but deposits its eggs either in a crevice of the cliff (it may be hundreds of feet above a boiling sea) or amongst the *débris* under the fallen rock-fragments at the foot of the cliffs. Sometimes they are laid under the large blocks of rock on the beach, and less frequently at considerable distances inland. Saxby states that he has found the eggs on grassy, rock-strewn slopes fifty or sixty yards from the sea; and Evans and Sturge found this bird breeding in Spitzbergen a mile or two inland. Sometimes the eggs can easily be obtained from the niche or crevice, but at others they are almost inaccessible. They are almost invariably two in number ; Macgillivray says that they are frequently three, and Audubon states that the latter number is the usual clutch."

Family PELECANIDÆ.

CORMORANT.

PHALACROCORAX CARBO, *Linn.*

Pl. XXXI., fig. 1.

Geogr. distr.—In suitable localities in Europe and Asia; ranging
to S. Africa, Australia, and New Zealand, and in America from
Hudson's Bay to Georgia. It occurs on most parts of the British
coasts, being, however, commoner in the north than in the south.
Food.—Fish and young birds.
Nest.—Of sticks and sea-weed, or, when in trees, a bulky structure
of sticks and twigs, lined with grass and weed.
Position of nest.—High up on cliffs and almost inaccessible rocks,
or in trees.
Number of eggs.—4.
Time of nidification.—IV-V; occasionally earlier or later.

Mr. Cecil Smith relates that Cormorants collect in
considerable numbers at their breeding stations, and that
they show a preference for fir-trees when not nesting upon
cliffs, in which case they sometimes occupy part of a
rookery.

Hewitson states that "Cormorants sometimes breed
upon the ledges of precipices, but choose, in preference,
those rocks which, standing isolated, are surrounded by
the sea, upon the tops of which they make their nests.
On the Fern Islands, where about forty or fifty pairs
breed, they occupy a low, flat island, slightly elevated
above the water, and confine themselves to one particular
and very limited part of it."

"The nests, which are placed at short distances from
each other, are large, and sometimes singularly lofty,
measuring upwards of two feet in height. They are
composed of a large quantity of the coarser sea-weeds, and
lined with finer weeds and dry grass."—(Ill. Eggs Brit.
Birds, vol. ii., pp. 415, 416.)

SHAG.

Phalacrocorax graculus, Linn.

Pl. XXXI., fig. 3.

Geogr. distr.—N. and W. Europe and the Mediterranean; not further eastward than the Black Sea; local in Great Britain, and commoner in the north than in the south.
Food.—Fish.
Nest.—A clumsy structure of twigs, grass, sea-weed, heather, &c.; nearly flat, or with shallow central cavity.
Position of nest.—In crevices of rocks or upon inaccessible ledges on the coast.
Number of eggs.—2-4.
Time of nidification.—VI.

In its nesting habits the Shag is frequently gregarious, as many as thirty nests having been counted upon a single small isolated rock.

Mr. Seebohm remarks that, "unlike the Cormorant, the Shag is almost exclusively a marine species, and seldom wanders from the sea to inland fishing grounds. It loves best to frequent those parts of the coast that are rocky, especially if there be small islands and plenty of caves and fissures amongst the cliffs, in which it not only rears its young, but takes shelter. Sometimes it may be seen sitting on the shelves of the cliffs, or more often basking on some sea-girt rock, with wings outspread, as if it were drying them after its aquatic gambols. In most of its habits it very closely resembles the Cormorant."

In explanation of the fact that the Shag often breeds in colonies, Mr. Seebohm says "the Shag is, if possible, a cave-breeding bird, and wherever caves are to be found on the coast, especially if they are only accessible to a boat, any available ledge where a nest can be placed is occupied by a pair of Shags. Caves of this kind are comparatively rare, and are consequently much sought after by these birds. On rocky coasts where there are no caves the Shag is generally found breeding in isolated pairs on ledges, and it is only on rocks where suitable ledges are not easily found, or where the Shag is especially numerous, that it becomes gregarious."—(Hist. Brit. Birds, vol. iii., p. 657.)

GANNET.

SULA BASSANA, *Linn.*

Pl. XXXIII., fig. 3.

Geogr. distr.—N. Europe, straggling southwards to the coasts of Africa. In Great Britain it is tolerably common, breeding on the Bass Rock, St. Kilda, Ailsa Craig, Lundy Island, Skellig Islands, the coast of Kerry, and the Stack of Suliskerry near the Orkneys.

Food.—Fish.

Nest.—Of grass and seaweeds in the shape of a flattened cone with a small central depression ; about 20 inches in diameter at its base, and from 6 to 8 inches high.

Position of nest.—Usually in the most inaccessible parts of the highest rocks.

Number of eggs.—1.

Time of nidification.—V-VI.

Mr. E. T. Booth, in his ' Rough Notes ' (Part V.), makes the following observations on this species :—" The note of the Gannet is powerful, though far from musical. If interfered with while sitting on the nest (and they seldom make a move unless threatened), the old birds will strike at the aggressor with their sharp-pointed bills, giving vent at the same time to a succession of hoarse, croaking sounds. Before daybreak I have, on two or three occasions, climbed to the summit of the Bass, and looked down on the silent multitudes collected on the ledges, while the first rays of the rising sun lit up the scene. In almost every instance the male and female were sitting side by side on the nest, the young, if small, being hidden from view, and those of larger size in most instances snugly nestled between the parents. As the daylight increases, first one and then another stretch out their necks, and, uttering a low note, rise up and flap their wings. It is soon an animated sight ; the old birds may be seen on all sides rubbing their heads together and going through the most amusing antics, the larger nestlings frequently thrusting up their heads between the pair and joining in the performance."

According to Mr. Howard Saunders an egg of the Gannet has been found as early as March, though, as a rule, incubation does not commence until the early part of May.

FAMILY LARIDÆ.

SUB-FAMILY *STERNINÆ.*

COMMON TERN.

STERNA FLUVIATILIS, *Naum.*

Pl. XXXII., figs. 1, 2.

Geogr. distr.—Europe generally, Asia Minor, probably generally in India as far as China, but rare in Central and Southern India; it has been found on the shores of the Red Sea, and, in America, in Labrador and Texas.

Food.—Fish, Crustacea, and insects.

Nest.—None.

Position in which eggs are deposited.—On flat, sandy or shingly shores, on mud banks or hillocks covered with short grass; sometimes near inland lakes.

Number of eggs.—3-4; usually 3.

Time of nidification.—V-VII.

As the eggs of the Common and Arctic Terns exhibit similar variations, so that the most that can be said to distinguish them is that those of the Common Tern are, as a rule, a little larger, and perhaps more frequently exhibit richly coloured varieties than those of *Sterna macrura*, it is important to be able to distinguish the two birds, which, though very similar, differ as follows:—The present species has a longer and stouter bill, of a red colour with a black tip; the black cap extends a little further and is of a more pointed form, the tail is longer than the closed wings, the under parts are white, and the tarsus is longer: in the Arctic Tern, on the other hand, the bill is entirely red in old birds, the black cap is of a more rounded form, the tail is not longer than the closed wings, and the under parts are greyish.

Mr. Cecil Smith says that Common Terns, "wherever their breeding station is, do not appear to trouble themselves much with making a nest, the eggs being placed in a hollow on the bare ground, or occasionally amongst the pebbles on the beach just above high water mark, without any hollow being made for them; occasionally also they are placed in hollows in the sand caused by drift sea-weed. The nests—if nests they can be called—are sometimes made in marshy places or on the borders of inland lakes."—('Birds of Somersetshire,' pp. 568-9.)

ARCTIC TERN.

STERNA MACRURA, *Naum.*

Pl. XXXII., figs. 4, 5.

Geogr. distr.—Northern Palæarctic and Nearctic regions, wintering as far south as the Cape of Good Hope; common in Great Britain, especially in the north.

Food.—Fish, Crustacea, and insects.

Nest.—A mere depression scratched in sand or grass, and lined with dry grass, straws, or feathers.

Position of nest.—Chiefly on small islands near the coast and sandy reaches on the coast.

Number of eggs.—2-4; usually 3.

Time of nidification.—VI.

The eggs exhibit all the variations found in the Common Tern, but are usually rather smaller. Dr. Saxby, speaking of the nesting of the Arctic Tern in Shetland, says:— " The eggs are usually deposited on a sandy or gravelly beach, or on a ledge of rugged bank which has been broken by the winter gales: in such places the eggs are merely laid in a hollow scraped out by the bird; but if the soil of the bank happens to be wet a small quantity of gravel is sometimes interposed. Often, however, the eggs are laid amongst the short grass further inland, and then the hollow is found to contain a few pieces of dead weeds or dry grass by way of lining."—(' Zoologist,' 1864, p. 9312.)

Mr. Howard Saunders states that, " As regards the British Islands, the Arctic Tern is the only species found breeding in the Shetlands, and it is by far the most abundant in the Orkneys, the Hebrides, and on the entire coast of Scotland. In England it breeds in numbers on the Farne Islands, and sparingly on the coast near the mouth of the Humber, south of which it has not yet been proved to nest on the east side of the island, nor along the shingly coast of Kent and Sussex, where the Common Tern occurs. Mr. Cecil Smith, however, states (Zool. 1883, p. 454) that he found it breeding on the Chesil beach in Dorsetshire. On the west side it breeds on the shores of Cumberland, on Walney Island in Lancashire, and probably on the Skerries and some other islands belonging to Wales; and Mr. Rodd states that it is far more abundant in the Scilly Islands than the Common Tern. In Ireland it has many breeding stations, from the Copelands, off Belfast, to the myriad islets of Galway and Kerry, and there are probably some on the eastern side of the island."—(Yarrell's Hist. Brit. Birds, 4th ed., vol. iii., p. 554.)

SANDWICH TERN.
STERNA CANTIACA, *Gmelin*.
Pl. XXXII., fig. 3.

Geogr. distr.—Europe and the coasts of Asia Minor; Africa south-
wards to the Cape; America southwards to Brazil; breeds locally on
the British coasts.
Food.—Fish.
Nest.—A mere shallow depression in the ground, lined with a few
blades of dry grass.
Position of nest.—Either on a flat mud bank or upon growing grass,
and amongst bushes on sand.
Number of eggs.—2-3; usually 2.
Time of nidification.—V-VI; June.

The Sandwich Tern is not easily frightened; it is said
that even the report of a gun does not appear to terrify it.
It is a regular summer visitor to our shores, but it is more
local than formerly. Mr. Seebohm says that " it no longer
breeds on the coast of Essex or Kent, but it is still found
in some numbers on the Farne Islands off the Northumberland coast, and there is a small colony on the coast of
Cumberland. A few pairs breed on Walney Island in
Lancashire, and on the Scilly Islands. In Scotland the
breeding-places of this handsome bird have fared no better.
It is carefully preserved on Loch Lomond, and is said still
to breed in the Frith of Tay, and in Sutherlandshire and
some other places on the west coast. In Ireland it breeds
in County Mayo, and possibly on some other parts of the
west coast."—(Hist. Brit. Birds, vol. iii., p. 272.)
Mr. Seebohm further relates that on the 19th June, 1870,
he took 149 eggs of this species on the Farne Islands,
selecting only such as were exceptionally handsome; the
eggs were in such numbers that it was impossible to walk
across the colony without treading upon some of them, and
after he had made his selection the numbers left in the
nests were not perceptibly lessened.

ROSEATE TERN.
STERNA DOUGALII, *Mont.*
Pl. XXXII., fig. 6.

Geogr. distr.—Europe from the British Isles to the Mediterranean; Ceylon; Andaman Islands; Cape of Good Hope; eastern coast of America; Australia. In Great Britain, according to Dresser, it breeds in Scotland and Ireland, and occurs rarely on the shores of England.
Food.—Fish.
Nest.—A slight depression surrounded by a circle of grass: according to Hewitson the eggs are sometimes "deposited upon a small quantity of dry grass."
Position of nest.—Usually upon small islands near the coast.
Number of eggs.—2-3.
Time of nidification.—V.

It seems doubtful whether this species breeds with any regularity in the British Islands. On the Farne Islands, where it was formerly abundant, several pairs were seen and shot as late as May, 1880. Hewitson, in 1846, mentioned that Mr. J. Hancock found it breeding in numbers upon Foulney Island on the Lancashire coast; and Mr. Howard Saunders says that in May, 1864 and 1865, Mr. Harting and he observed it on the neighbouring Walney Island. He adds:—"It probably nests in a few localities on the coast of Scotland, but statements regarding its breeding on Loch Lomond or any other lochs appear to be devoid of foundation. So far as is known the Roseate Tern nests almost exclusively on islands, and generally on those which are remote and storm-beaten. Off the coast of Ireland, where there are many such islets, several breeding places have been enumerated by Thompson; but most of these have since been abandoned, and although the birds have probably migrated to other and less disturbed localities, it would not be easy, even if it were desirable, to indicate precisely the places where colonies may still be found."— (Yarrell's Hist. Brit. Birds, vol. iii., p. 545.)

Mr. Seebohm says "it is doubtful whether the Roseate Tern breeds in any part of the British Islands at the present time."

LITTLE (OR LESSER) TERN.
STERNA MINUTA, *Linn.*
Pl. XXXII., fig. 8.

Geogr. distr.—Temperate Europe; wintering on the western coast of Africa, and as far south as the Cape; also found in Western America. In Great Britain it is scattered here and there from the southern coast of England to the Orkneys, but is most common along the eastern coasts.

Food.—Aquatic insects and small fish.

Nest.—A mere hollow scooped out of the shingle or mud.

Position of nest.—Usually in gravelly sand on the sea-shore.

Number of eggs.—8.

Time of nidification.—V-VI.

In his 'Birds of the West of Scotland,' p. 471, Mr. Robert Gray says of the Little Tern :—"In its habits it resembles other Terns, but is even a more interesting study to the ornithologist, being so much smaller than the Common or Arctic Terns. During the breeding time it shows considerable impatience on its haunts being invaded, and meets the intruder with shrill cries, which it persistently utters so long as he remains in the neighbourhood. On dispersion of these breeding colonies it travels by slow degrees southwards, and finally quits our shores in September, arriving again in the following spring."

Although the above account would give one the idea that the Little (or Lesser) Tern was much alarmed when disturbed from the nest, such does not appear to be the case. Mr. Howard Saunders says that he "found considerable numbers of this Tern at the mouth of the Thames, on the Kentish side, about Yantlet Island and the creek of the same name close by;" and he adds, "When their breeding haunts are visited they exhibit but little fear, settling on the ground not far from those who may be looking for their eggs or young, and will frequently walk about with a light step, or, with a piping note, again take wing."—(Yarrell's Hist. Brit. Birds, 4th ed., vol. iii., p. 558.)

The bird breeds pretty generally along the coasts of Scotland and Ireland, and in suitable places in Wales. In England it nests in Lancashire, Cumberland, Yorkshire, Lincolnshire, Norfolk, Suffolk, Essex, and perhaps Dorsetshire.

BLACK TERN.
HYDROCHELIDON NIGRA, *Linn.*
Pl. XXXII., fig. 7.

Geogr. distr.—Europe generally, northward to Scandinavia; eastward in Asia to Turkestan; during the winter some distance southward in Africa; in America from Canada to Chili and Peru. In Great Britain it is rare; it formerly used to breed in Cambridge, Lincoln, Kent, and Norfolk, but it is doubtful whether it continues so to do.

Food.—Fish, Crustacea, and insects.

Nest.—Tolerably well constructed of coarse grass, flags, &c.

Position of nest.—Upon tufts of grass or rushes in wet marshy places, and near inland pools of water.

Number of eggs.—3.

Time of nidification.—V.

Mr. Gray says that "of late years Black Terns have been observed in the spring time and autumn in many Scottish counties; but these, generally speaking, have been stray birds. In Haddington, Berwick, Aberdeen, Fife, and Dumfries-shire, many specimens have from time to time been shot and preserved.

"In the west of Scotland small flocks occasionally appear at Loch Fyne and other sea reaches. Mr. George Hamilton informs me that he and his brother observed five specimens near Minard in September, 1860; and I have myself seen the species on Loch Lomond, flapping round the boat in which I was rowing, within a distance of eight or ten yards."—('Birds of the West of Scotland,' p. 474.)

Mr. Gray goes on to mention other specimens shot in Scotland as late as 1870.

Though formerly abundant as a breeding species in Great Britain, the Black Tern can now only be regarded as a spring and autumn visitor to our country. According to Mr. Stevenson, the last nest known to him in Norfolk was discovered by a marshman at Sutton in 1858, who, with characteristic humanity, shot the birds.

Sub-family *LARINÆ.*

BLACK-HEADED GULL.

Larus ridibundus, *Linn.*

Pl. XXXIII., figs. 1, 2.

Geogr. distr.—Western Europe to Eastern Asia and Scandinavia to Northern Africa, in suitable localities. In Great Britain it is common, widely distributed, and resident, breeding in Northumberland, Durham, Lincolnshire, Norfolk, Essex and Kent; also abundantly in Ireland and Scotland.

Food.—Small fish, frogs, insects in all stages, and worms.

Nest.—Fairly well made of reeds and dried grass.

Position of nest.—On the ground or on beaten-down rushes on islets well surrounded by water in inland lakes and marshes.

Number of eggs.—3-4; usually 3.

Time of nidification.—IV-V.

Mr. Seebohm tells us that "the most celebrated breeding place of the Black-headed Gull in the British Islands is Scoulton Mere, not far from Higham in Norfolk. The lake is entirely surrounded by plantations of oak, beech, Scotch fir, and spruce-fir. It covers about 150 acres, but 70 acres of this area are taken up with a large island upon which the gullery is situated. The colony consists of about 8000 birds, and is said to be gradually increasing in size."

"The part of the island where the Gulls breed consists of a few acres of swampy ground, thinly sprinkled over with flags and coarse grass, in which the nests are placed, and planted here and there with clumps of low birches and willows. When I visited this gullery in company with Mr. Bidwell on the evening of the 13th of last May (1884), the swamp was crowded with birds, which looked very conspicuous amongst the flag and sedge. As we neared the island thousands rose from the ground, and before we landed, the air was one mass of birds, wheeling round and round in interlacing circles, whilst their cries were incessant. After selecting a few eggs we recrossed to the margin of the lake, and watched the seething mass of birds. It was a most animating sight."—(Hist. Brit. Birds, vol. iii., pp. 311, 312.)

COMMON GULL.

LARUS CANUS, *Linn.*

Pl. XXXIII., fig. 5.

Geogr. distr.—Europe generally; northward to the North Cape of Scandinavia, being common in Russia and Finland; eastward throughout Siberia and southward to China; occurring also in Malta, Greece, and the coast of Asia Minor; also in Africa from Egypt westward to Algeria, the Canaries, and Madeira. In Great Britain more common in winter than in summer, but breeds in the Outer Hebrides and the Shetland Islands, and has been recorded as breeding in the south of England.

Food.—Fish and Crustacea.

Nest.—Formed of fuci, occasionally grass, bits of turf, ling, dry sea-weed, and other vegetable substances.

Position of nest.—On drift left by the tide, on green turf, on tussocks on marshy ground, on the shores of lakes, or in old Crows' nests on the top of fir trees; also on the face or summit of a precipitous cliff amongst sedgy grass, green samphire or ling, or even on the bare rock.

Number of eggs.—3.

Time of nidification.—V-VI.

The following observations are extracted from Mr. E. T. Booth's ' Rough Notes ' (Part VII.) :—" Though Common Gulls are by no means so ready as many other species to attack any feathered stranger that ventures near their breeding-quarters, I witnessed an amusing scene at Loch Inver, in Sutherland, one evening in June, 1877. Several pairs of Gulls frequent the lonely rock-bound coast to the south-west of the loch during summer ; one nest, however, was placed just above the wash of the tide, but a short distance from the village. While watching the two old birds fishing along a sandy bay in the immediate vicinity of the low-lying ridge on which their young were located, I noticed a Long-eared Owl flap slowly towards the water. Evidently disturbed from its shelter in the pine woods before the accustomed hour, dazzled by the light, and apparently at a loss which way to turn, its uncertain and wavering flight speedily attracted attention. Instantly the male Gull, with loud screams, dashed after the intruder, and, buffeting the bewildered bird repeatedly, forced it out to sea ; roused by the outcry, a fresh contingent of Gulls shortly arrived, and at once joined in the attack with the greatest fury. The Owl, after having been driven over the centre of the loch, at length rose high in the air,

o

followed by a couple of the most inveterate of its pursuers, and not till it had disappeared from view among the hills to the north did the Gulls return to their quarters and was peace re-established. On no other occasion have I seen an Owl rise to such an altitude; at one time it wheeled round in large circles at the height of at least three hundred feet."

LESSER BLACK-BACKED GULL.
Larus fuscus, *Linn.*
Pl. XXXIV., fig. 1.

Geogr. distr.—Throughout N. Europe in summer; eastward to China and Dauria and westward to the Canaries, in winter straggling down to N. Africa. In Great Britain it breeds commonly on the coast, but more especially in the north; being more numerous in Scotland than in England.

Food.—Fish, Crustacea, worms, and insects in all stages; also eggs of other rock-frequenting birds.

Nest.—A depression lined with leaves, dry grass, and mosses, rarely mixed with fragments of sea-weed.

Position of nest.—On bare and barren islands, or near inland lakes in a tuft of rushes, or amongst heather and coarse herbage.

Number of eggs.—2-3.

Time of nidification.—VI.

It is supposed that this Gull hunts by scent, as, when in search of food, it usually flies up-wind. Mr. Booth, however, makes some remarks tending to show that this is by no means invariably the case; he says:—"When the shoals of mackerel arrive off the south coast in the spring, hundreds of seine-boats are engaged in watching for the fish to come to the surface; as soon as they are sighted the crews row rapidly to the spot, and, shooting the net round them, frequently enclose large numbers. Should any Gulls, however, be near at hand, their sharp eyes are sure to detect the first ripples on the water, and, dashing down into the middle of the shoal, the fish are driven to the bottom, and the men, who have rowed hard for half-a-mile or more, and possibly paid out a portion of their net, find their time and labour thrown away, while the mischievous birds, with a derisive scream, sail off to repeat the performance at the earliest opportunity. While watching these proceedings off Brighton and Shoreham, I have often been requested by the crews of the boats to shoot the Gulls, the men declaring that, what with the Bird Act and the Gun License they were unable to help themselves, being forced to stand quietly by while the birds snatched the bread from their mouths."—('Rough Notes,' Part IX.)

GREATER BLACK-BACKED GULL.

LARUS MARINUS, **Linn.**

Pl. XXXIV., fig. 2.

Geogr. distr.—In Europe almost restricted to the northern and central parts, but straggling southwards in winter; N.E. America. In Great Britain it is generally distributed on the coast during the winter, but in the summer it principally frequents the northern counties, breeding, however, on the coast of Caermarthenshire, the Steep Holmes, and Lundy Island, but most abundantly in the Orkney and Shetland Islands, &c.

Food.—Fish and Crustacea.

Nest.—Bulky, formed of dry grass, moss, heather, sheep's wool, and large feathers carelessly heaped together.

Position of nest.—On the ground amongst grass.

Number of eggs.—4.

Time of nidification.—V-VI.

Mr. Booth says that "in the Highlands the Greater Black-Backed Gull causes considerable loss to many of the small sheep farmers and crofters, who are unable to give the necessary care and protection to the few animals they possess. A weakly ewe is no sooner discovered than she is set upon, and after being forced into some crevice among the rocks, or slowly butchered by thrusts from their power-ful bills, the lamb next falls an easy victim. Such facts, I am aware, have been denied by some writers; but during the last few years several instances have come under my observation, in addition to the reports I have heard from shepherds and small owners. The young of Grouse and many other birds breeding on the moors are also greedily devoured by these robbers, and no exposed egg is safe if once it has attracted their notice.

"The nest, like that of most sea-birds, is by no means elaborately constructed, differing slightly in its composition according to the locality. On the sea-coast it is placed among the rough herbage on the upper ledge of a rocky cliff, or at times in some cavity among the bare stones. Coarse grass with strands from any adjacent plants are the principal materials, with now and then a small quantity of fine dead sea-weed. A few feathers are usually to be seen in the nest or scattered around, but these are probably plucked by the birds themselves while cleaning their plumage."—('Rough Notes,' Part IX.)

HERRING GULL.

LARUS ARGENTATUS, *Gmelin*.

Pl. XXXIV., fig. 3.

Geogr. distr.—Throughout northern and central Europe eastward to Russia, and in the breeding season throughout Europe; N.E. America. Resident and widely distributed in Great Britain, occurring on most parts of our coasts during the breeding season.

Food.—Fish, Mollusca, Crustacea, and starfish.

Nest.—Usually bulky and formed of rough dry grass, with part of the sod attached, and fragments of sea-weed, or sometimes a mere hole scratched in the soil, with little or no lining.

Position of nest.—Usually on patches of grass on precipitous rocks, or on ledges of cliffs, but sometimes on the surface of low, flat rocks amongst grass and loose stones.

Number of eggs.—2-3.

Time of nidification.—V.

This species has been frequently known to breed in confinement in a garden. In a wild state it is shy and suspicious, and very often, by its angry cries, gives notice to other birds in its vicinity of the approach of an invader. Its flight is heavy and laboured, but its walk, though somewhat laughable, is in a measure graceful; the steps taken deliberately; but, with the least excitement, changing into the most ridiculous little tripping run, reminding one of a delicate lady crossing a crowded thoroughfare.

Mr. Harting says :—" The cry of the Herring Gull is not unlike that of the Common Gull, a sort of hoarse laugh or a cackle, sounding like '*wa-agh-agh-agh-agh.*' Sometimes a barking cry is preceded by a prolonged squeal, like '*whee-e-e-kiark-kiark-kiark-kiark,*' and is generally uttered when they are frightened from the nest."—(' Sketches of Bird Life,' p. 286.)

Mr. Booth (' Rough Notes,' Part VII.) expresses a conviction that the farmer, rather than the game-preserver, suffers from the attacks of this bird, which, when food is scarce, shows a great liking for turnip roots, and makes great havoc with newly-sown grain. In the Fern Islands, however, in company with the Lesser Black-Backed, it destroys great quantities of eggs of the Eider Duck.

KITTIWAKE.

RISSA TRIDACTYLA, *Linn.*

Pl. XXXIII., fig. 4.

Geogr. distr.—Northern portions of Old and New World ; travelling southwards in the winter, when it strays into the middle States of N. America ; rare in the Mediterranean. In Great Britain it is resident and generally distributed, breeding in great numbers at Flamborough Head, Bass Island, and rocks near the Castle of Slains in Aberdeen, and on Priestholm Island.

Food.—Fish and Crustacea and other marine animals.

Nest.—Bulky, formed of dry grasses, sea-weed, and clay.

Position of nest.—On jutting ledges and projections of high and precipitous cliffs, washed or surrounded by the sea.

Number of eggs.—2-3 ; rarely 4.

Time of nidification.—V.-VI.

Mr. Robert Gray says that "on Ailsa Craig the Kitti-wakes are among the first birds to arrive, and for a day or two during the time of nest-building they are seen tearing up the loose turf,—the clamour of the birds while at this employment being almost as bewildering as when they are pursuing their prey. Some of the nests, the foundations of which are laid with turf with the loose earth adhering to it, are placed on the upper ledges at the elevation of 500 or 600 feet, while others are quite within reach of the visitor as he passes along the rough road at the foot of the cliffs. In course of time the bottom of the nest, through rain and spray, becomes trampled into a kind of clay, which looks as if the nest had originally been built of mud, and hence the inaccurate reports of some observers. While incubating, these gentle birds are tame and confiding, seldom taking wing if fired at or otherwise disturbed, but should one or two be fired at and fall back dead on the nest, the neigh-bours will then rise on wing and flit about, making pitiable lamentation, and crying all the while *Kittawee, Kittawee ! Ah, get away, get away !*

" I hope that no true ornithologist or sportsman will find fault with me for saying that to practise this kind of shoot-ing is a shame."—(' Birds of the West of Scotland,' p. 479.)

SUB-FAMILY *STERCORARIINÆ.*
COMMON SKUA.
STERCORARIUS CATARRHACTES, *Linn.*

Pl. XXXV., fig. 1.

Geogr. distr.—Northern parts of the Atlantic and North Sea; occurring rarely in the mainland of Europe, and principally on the western and north-western coasts. In Great Britain during the breeding season it appears to be restricted to the Shetland Isles; occurs on the eastern and rarely on the western coasts of England, and occasionally on the coast of Ireland.

Food.—Mollusca, Crustacea, fish, and young birds.

Nest.—A considerable depression in mossy ground, lined with a quantity of moss and dried grass, forming a nest about twelve inches in diameter by three in depth.

Position of nest.—On elevated moorland.

Number of eggs.—2.

Time of nidification.—VI.

The Rev. C. A. Johns, in his ' British Birds in their Haunts,' says that the Common Skua will attack even the Eagle if he approaches its nest. He adds:—" I once saw a pair completely beat off a large Eagle from their breeding place, on Rona's Hill."—(P. 591.)

Mr. Robert Gray says:—" The fact of the breeding of this Skua being strictly confined to the Shetland Islands has of late years led to so much destructive intrusion by collectors that it is now only by the most careful protection that the birds are enabled to maintain a footing there. Thirty years ago there were three separate nesting localities, *viz.*, the outlying islands of Foula and Uist, and Rona's Hill on the main island. The last-named haunt is now entirely deserted, and in the two others the number of Skuas which yearly resort thither for nesting purposes is comparatively small. From these haunts a few usually find their way southwards along the coasts of the eastern counties ; but the bird is of rare occurrence in the west.

" During the breeding season the Common Skua becomes quite fearless, attacking any intruder * in its haunts with so much spirit as occasionally to drive both man and dogs off the ground. Its habits at other seasons can seldom be observed, as it does not often come near the shore."— (' Birds of the West of Scotland,' p. 493.)

---

* The attack is directed at his eyes, according to Mr. Gray.

RICHARDSON'S SKUA.

STERCORARIUS CREPIDATUS, *Banks.*

Pl. XXXV., fig. 2.

Geogr. distr.—Northern portions of Europe and America ; in Asia on the coast of Hindostan ; in Africa southwards to the Cape. In Great Britain it is not uncommon on different parts of the coast in winter, but it breeds only in the Orkney and Shetland Islands and the Hebrides.

Food.—Fish, Mollusca, and eggs of sea birds.

Nest.—Constructed rather carelessly of grass, moss, and fragments of heather.

Position of nest.—On the ground amongst heather in marshy and uncultivated moorland.

Number of eggs.—2.

Time of nidification.—V-VI.

This bird has been met with as far south as Yorkshire. It is a notorious bandit, robbing Gulls and Terns of their hard-earned meals. Mr. Cecil Smith says of it, " This, like the rest of the Skuas, is a northern species, only visiting our southern counties in the autumn and winter, probably following the other birds that follow the herrings and sprats. It remains, however, to breed in the more northern parts of the kingdom, such as the Shetlands and Orkneys, where it appears to occupy very high hills and moors as its breeding station, scratching a hole amongst the heather for a nest, which it lines with dry grass and moss."

Mr. Smith further remarks that this Skua attempts to mislead the birds-nester by feigning to be wounded, after the manner of the Peewit. Its food appears to consist mostly of fish, which it bullies the Gulls and Terns into disgorging, but it is also a bold thief, losing no opportunity of plundering the nests of any Gulls, Guillemots, or other birds which are unlucky enough to be near neighbours during the breeding season.—(' Birds of Somersetshire,' pp. 617, 618).

BUFFON'S SKUA.
STERCORARIUS PARASITICUS, Linn.
Pl. XXXV., figs. 3, 4.

Geogr. distr.—Northern Palæarctic and Nearctic Regions. In Great Britain it occurs sparingly, being rare on the southern coast of England; it is, however, said to breed in Scotland.
Food.—Crowberries, insects, Crustacea, fish, small birds, and carrion.
Nest.—A mere depression in the ground.
Position of nest.—In mossy, heathy, or grassy wastes.
Number of eggs.—2-3; usually 2.
Time of nidification.—VI.

Mr. Gray says that " on the mainland of the west of Scotland this Skua is only a straggler ; but it is probably a regular summer visitant to the outer islands. A specimen was shot in Skye, in the autumn of 1855, and exhibited at a meeting of the Royal Physical Society of Edinburgh ; and in the summer of 1863 I examined a pair that were shot on the island of Wiay by Colin McRuig, Esq., surgeon, as they hovered over a marsh where there were nests of Richardson's Skua and other birds. The likelihood is they had a nest on the spot." " Buffon's Skua has been found breeding in Caithness-shire, though not for some years past ; and also in Shetland, as I have been informed by Mr. Dunn, who procured the eggs from one of three nests in the island of Hoy, fifteen years ago."—(' Birds of the West of Scotland,' pp. 498, 499.)

In Mr. Seebohm's opinion, " the statements that this bird has bred in the Hebrides, in Caithness, and in the Orkneys, are founded upon very meagre and entirely insufficient evidence."

Like the other species of Skua, this bird is very valiant in defence of its nest, flying with the utmost fierceness at the intruder, who has no little difficulty in defending himself against its attacks.

POMATORHINE (OR POMARINE) SKUA.
STERCORARIUS POMATORHINUS, *Temm.*

Pl. XXXV., fig. 5.

Geogr. distr.—N. Europe, Asia and America, being rare in the central and southern parts; in Africa on the western coast as far south as Walwich Bay, and in N. Australia. A rare winter visitant in Great Britain; in Scotland more frequent on the eastern than western shores. Thousands were seen, in the autumn of 1880, off the York-shire coast.

Food.—Crustacea, fish, and young birds.

Nest.—A mere depression in the ground.

Position of nest.—In moss-covered moorland.

Number of eggs.—2.

Time of nidification.—VI.

This bird has been shot on the coasts of Lancashire, Kent, Sussex, Hampshire, Devon, and Cornwall, and other parts of the English shores; it appears to be most rare on the eastern coast. According to Mr. Booth ('Rough Notes,' Part VII.), "its flight is more steady than that of the Arctic Skua, and while on passage it keeps at a greater elevation." Speaking of a pair which he kept in confinement, he says : —"The captives thrived well on herrings, mackerel, or sprats, their actions, while feeding, being exceedingly singular. If one happened to seize a portion of food too large to be swallowed with ease, he would call loudly, when his companion at once came rapidly up, and, clutching hold of one end of the fish, each would tug lustily till the whole was divided, when the parts were consumed by the pair in the most amicable manner. This curious per-formance was now and then repeated half-a-dozen times during the same meal."

Mr. Gray observes that "this species of Skua does not entirely depend upon its piratical exertions for subsistence, but contents itself occasionally with a diet of putrid fish or small dead animals which it happens to meet with in its flights along the shore. It has even been known to devour rats and birds."—('Birds of the West of Scotland,' p. 495.)

Family PROCELLARIIDÆ.

MANX SHEARWATER.

Puffinus anglorum, *Temm.*

Pl. XXXVI., fig. 1.

Geogr. distr.—Throughout the N. Atlantic ocean, not extending into the Baltic, but into the Mediterranean as far as the Black Sea; on the American coast from Labrador to New Jersey; also in Bermuda: in Great Britain it is not uncommon, especially on the western coasts, breeding on the Scilly Islands, Lundy, Staffa, the Outer Hebrides, Orkney and Shetland Islands.

Food.—Fish, Mollusca, Crustacea, and worms.

Nest.—A few straws or dry plants at the extremity of a burrow, one, two, or more feet deep.

Position of nest.—In the sandy soil of steep cliffs.

Number of eggs, 1.

Time of nidification.—V-VI.

According to Dixon, in Mr. Seebohm's ' History of British Birds," the cry of this bird may be expressed on paper as " *kitty-coo-roo, kitty-coo-roo.*" He adds :—" This note is uttered both when the bird is on the wing and when sitting on its nest. Guided by the note, the islanders are able to find the nests with little difficulty, so that they always prefer to go in search of this species at night. Dogs are also trained for the purpose of finding the holes."

" The Shearwater burrows in the ground like a Puffin or a Petrel, and the holes are sometimes very long, and often under large masses of rock, where it is impossible to reach the nest."—(Vol. iii., p. 422.)

Mr. Gray observes that " there are numerous breeding haunts of this Shearwater throughout the West of Scotland, and the bird itself may be called abundant within the circle of the Inner Hebrides." He says that westward of that group the only breeding localities known to him are Pabbay—one of the isles of Barra, and St. Kilda. To the above haunts he adds—the island of Eigg and the Treshinish isles, Staffa, Iona, and various small rocky islets. The Shearwater appears in April and remains until October.—' Birds of the West of Scotland,' pp. 503-4.)

FULMAR.

Fulmarus glacialis, *Linn.*

Pl. XXXVI., fig. 3.

Geogr. distr.—Shores of N. Europe and America, breeding in the high north, migrating southwards in winter : in Great Britain it is rare, but more so in Ireland than England; has been obtained off the coasts of Wales and Cornwall, Essex, Yarmouth Roads, Flamborough Head in Northumberland, and Durham : it breeds on the islands off the Scotch coast, especially St. Kilda.

Food.—Blubber, fish, Mollusca, and barnacles.

Nest.—Usually a shallow excavation in turf, lined with dry grass and sea-pink.

Position of nest.—Generally on almost inaccessible rocks skirting the ocean, one or more nest being placed upon grass-grown ledges.

Number of eggs.—1.

Time of nidification.—VI.

Mr. Cecil Smith says that the Fulmar " is not uncommon on the more northern coast of England and Scotland, and breeds in some of the islands off the coast, making its nest on the grassy shelves of the highest precipices. The nest itself is formed of herbage, seldom bulky, generally a mere shallow excavation in the turf, lined with dry grass and the withered tufts of the sea-pink."—(' Birds of Somersetshire,' p. 422.

Mr. R. Gray states that the breeding-quarters of the Fulmar are St. Kilda, Soa, and Borrera, from which group of rocks it is a straggler in the summer to the Outer Hebrides. In the Shetlands it was only known as a visitor until the 4th of June, 1878, when about a dozen pairs were observed hovering round the cliffs of the island of Foula, where they reared their young in some places in which, according to the natives, no birds had ever bred before. The nests were placed on small ledges formed by the splitting of the rocks into layers. Since that year the species is believed to have increased on the island.—(See ' Zoologist,' 1879, p. 380.)

LEACH'S (or FORK-TAILED) PETREL.

Procellaria leucorrhoa, *Vieill.*

Pl. XXXVI., fig. 4.

Geogr. distr.—Atlantic ocean from St. Kilda southward to Madeira and on the American coast from Labrador southwards to Washington; also on the western coast of N. America: in Great Britain it is not rare, especially on the western coast of England, but chiefly in November and December; on the eastern coast of Scotland it is extremely rare: it has been recorded from Ireland; it formerly bred in the Orkneys, but is now only known to breed at St. Kilda.

Food.—Fish, Mollusca, and Crustacea.

Nest.—Flat, formed of fine grasses carelessly put together, and occasionally a few pebbles, at the end of a burrow two to three feet deep, like that of the Sand-Martin.

Position of nest.—Chiefly on grassy islands, the burrow being made under the sod and often under rocks.

Number of eggs.—1.

Time of nidification.—VI.

Mr. Cecil Smith says:—"The food of this little Petrel consists chiefly of Mollusca, small fish, and Crustacea, which it picks up amongst floating sea-weed, and of any greasy substances which are found around fishing boats or ships out at sea.

"The Fork-tailed Petrel breeds in sandy burrows or in holes in rocks."—('Birds of Somersetshire,' p. 625.)

According to Mr. R. Gray, the Fork-tailed Petrel, which was discovered in the island of St. Kilda about fifty years ago, has since been found breeding there in a colony which has established itself on Dun,—an isolated stack,—under the loose rocks near the summit. It is also known to frequent the island of Mingalay, in Barra, where a few pairs incubate every year in company with the Storm Petrel.—(See 'Birds of the West of Scotland, p. 505).

Audubon says that when incubating these Petrels remain in their burrows until towards sunset, when they start off in search of food, returning to their mates or young in the morning, and feeding them then.

STORM PETREL.

PROCELLARIA PELAGICA, Linn.

Pl. XXXVI., fig. 5.

Geogr. distr.—Atlantic ocean generally; also the eastern, western, and south-western coasts of Africa : tolerably common off the British coasts, breeding on the Scilly Islands, the rocky shores of the north of Cornwall, and on the Shetland Islands and St. Kilda.

Food.—Small fishes, Mollusca, and Crustacea.

Nest.—Merely a collection of fragments of plants in a depression in the ground, or in holes in the soft mud, the entrance being as large as a rabbit-burrow, leading to branching galleries, in which several pairs breed in company.

Position of nest.—Amongst trees or in hollows in cliffs.

Number of eggs.—1.

Time of nidification.—VI-X.

Mr. Cecil Smith observes that this species "breeds in a hole or crevice of rocks or rabbit hole.' He adds :—"Mr Sanford gave me a curious description of a Storm Petrel breeding-station, which he had visited, on a small rocky island off the coast of Galway, called Hii Island. On the top of this island there is an ancient building, like the domed hut of an Esquimaux. The walls of this hut are very thick, nearly five feet, and in the holes in these walls the storm Petrel bred in considerable numbers, but on no other part of the island, neither in the crevices in the rocks nor in the holes in the ground."—'Birds of Somersetshire,' p. 627.)

According to Mr. Howard Saunders, this species breeds "freely at many different places, generally on small islands; but is never observed to frequent land except during the breeding season"—(Yarrell, 4th ed., vol. iv., p. 43.)

Hewitson says that the nests have the appearance of being carefully made, of small bits of stalks of plants and pieces of hard dry earth. He adds that "during the day the old birds remain within their holes, and, when the other birds are gone to rest, issue forth in great numbers, spreading themselves far over the surface of the sea."

ADDENDUM.

AQUATIC WARBLER.

ACROCEPHALUS AQUATICUS, *Gmelin*.

Pl. VIII., fig. 20.

The egg of this species was lent to me by inadvertence for that of the Marsh Warbler, and, although the error was soon discovered, it was unfortunately too late to prevent its appearance on the plate, which was by that time printed off.

Although no instance of the breeding of the Aquatic Warbler in Great Britain has been, to my knowledge, recorded, several specimens have been shot in England, one, at least, of which was obtained in the summer time.

Mr. Seebohm says:—"It is no subject for surprise to find this bird occasionally wandering across the English Channel, when we know it breeds pretty commonly on the opposite coasts of France and Holland."

The habits of this species resemble those of the Sedge rather than the Reed Warbler, and its egg is not unlike some of my varieties of the Sedge bird (*cf.* Pl. IX., fig. 4).

1886.

June . Part I., containing Sheets B, C, and Titling, Plates I.—VI. and frontispiece.

July . Part II., containing Sheets D, E, F, and Plates VII.—XII.

August. Part III., containing Sheets G, H, and Plates XIII.—XVIII.

Sept. . Part IV., containing Sheets I, K, L, and Plates XIX.—XXIV.

Oct. . Part V., containing Sheets M, N, and Plates XXV.—XXX.

Nov. . Part VI., containing Sheet O, List and Index, and Plates XXXI.—XXXVII.

ERRATA.

Page 8. The statement respecting the favourite food of the Merlin is said to be an error.

,, 15, line 28, *after* " near " *add* " Halberstadt."

,, 54, line 5, *for* " Pl. IX." *read* " Pl. X."

,, 56, line 10 from bottom, *for* " Cowfield " *read* " Cowfold."

,, 80, line 4, *for* " Pl. XII." *read* " Pl. XIII."

,, 133, *after* line 3, *insert* " Pl. XX., fig. 3."

A CLASSIFIED LIST

OF

BIRDS WHICH BREED OR HAVE UNTIL RECENTLY BRED IN GREAT BRITAIN.

———◆———

The subjoined list is arranged in accordance with Mr. Dresser's 'List of European Birds.'* It contains only such species as are recorded in the present volume, those which are not known to breed in the British Isles being excluded.

In this catalogue of Birds it has not been thought necessary to include the higher divisions, which (though interesting and instructive enough to the ornithologist) are apt to puzzle the young egg-collector: moreover, any enquirer after groups more comprehensive than families can easily obtain the European List.

———————

Family TURDIDÆ.

Sub-family *TURDINÆ*.

1. TURDUS VISCIVORUS, *Linn.*
   Missel Thrush.

2. TURDUS MUSICUS, *Linn.*
   Song Thrush.

3. TURDUS MERULA, *Linn.*
   Blackbird.

4. TURDUS TORQUATUS, *Linn.*
   Ring Ouzel.

Sub-family *CINCLINÆ*.

5. CINCLUS AQUATICUS, *Bechst.*
   Dipper.

Sub-family *SAXICOLINÆ*.

6. SAXICOLA ŒNANTHE, *Linn.*
   Wheatear.

7. PRATINCOLA RUBETRA, *Linn.*
   Whinchat.

8. PRATINCOLA RUBICOLA, *Linn.*
   Stonechat.

9. RUTICILLA PHŒNICURUS, *Linn.*
   Redstart.

10. RUTICILLA TITYS, *Scop.*
    Black Redstart.

Sub-family *SYLVIINÆ*.

11. ERITHACUS RUBECULA, *Linn.*
    Redbreast.

———————

* 8vo, 1881, price one shilling, published by the author, at 6, Tenterden Street, Hanover Square, W.

12. DAULIAS LUSCINIA, *Linn.*
Nightingale.

13. SYLVIA RUFA, *Bodd.*
Whitethroat.

14. SYLVIA CURRUCA, *Linn.*
Lesser Whitethroat.

15. SYLVIA ATRICAPILLA, *Linn.*
Blackcap.

16. SYLVIA SALICARIA, *Linn.*
Garden Warbler.

17. MELIZOPHILUS UNDATUS, *Bodd.*
Dartford Warbler.

Sub-family *PHYLLOSCOPINÆ.*

18. REGULUS CRISTATUS, *Koch.*
Golden-crested Wren.

19. PHYLLOSCOPUS COLLYBITA, *Vieill.*
Chiffchaff.

20. PHYLLOSCOPUS TROCHILUS, *Linn.*
Willow Wren.

21. PHYLLOSCOPUS SIBILATRIX, *Bechst.*
Wood Wren.

Sub-family *ACROCEPHALINÆ.*

22. ACROCEPHALUS STREPERUS, *Vieill.*
Reed Warbler.

23. ACROCEPHALUS PALUSTRIS, *Bechst.*
Marsh Warbler.

24. ACROCEPHALUS SCHÆNOBÆNUS, *Linn.*
Sedge Warbler.

25. LOCUSTELLA NÆVIA, *Bodd.*
Grasshopper Warbler.

Family ACCENTORIDÆ.

26. ACCENTOR MODULARIS, *Linn.*
Hedge Accentor.

Family PANURIDÆ.

27. PANURUS BIARMICUS, *Linn.*
Bearded Reedling.

Family PARIDÆ.

28. ACREDULA ROSEA, *Blyth.*
British Long-tailed Titmouse.

29. PARUS MAJOR, *Linn.*
Great Titmouse.

30. PARUS BRITANNICUS, *Sharpe & Dresser.*
British Coal Titmouse.

31. PARUS PALUSTRIS, *Linn.*
Marsh Titmouse.

32. PARUS CÆRULEUS, *Linn.*
Blue Titmouse.

33. LOPHOPHANES CRISTATUS, *Linn.*
Crested Titmouse.

Family SITTIDÆ.

34. SITTA CÆSIA, *Wolf.*
Nuthatch.

Family CERTHIIDÆ.

35. CERTHIA FAMILIARIS, *Linn.*
Common Creeper.

Family TROGLODYTIDÆ.

36. TROGLODYTES PARVULUS, *Koch.*
Common Wren.

Family MOTACILLIDÆ.

37. MOTACILLA ALBA, *Linn.*
White Wagtail.

38. MOTACILLA LUGUBRIS, *Temm.*
Pied Wagtail.

39. MOTACILLA FLAVA, *Temm.*
Blue-headed Wagtail.

40. MOTACILLA MELANOPS, *Pall.*
Grey Wagtail.

41. MOTACILLA RAII, *Bonap.*
Yellow Wagtail.

42. ANTHUS PRATENSIS, *Linn.*
Meadow Pipit.

43. ANTHUS TRIVIALIS, *Linn.*
Tree Pipit.

44. ANTHUS OBSCURUS, *Lath.*
Rock Pipit.

Family ORIOLIDÆ.

45. ORIOLUS GALBULA, *Linn.*
Golden Oriole.

Family LANIIDÆ.

46. LANIUS COLLURIO, *Linn.*
Red-backed Shrike.

Family MUSCICAPIDÆ.

47. MUSCICAPA GRISOLA, *Linn.*
Spotted Flycatcher.

48. MUSCICAPA ATRICAPILLA, *Linn.*
Pied Flycatcher.

Family HIRUNDINIDÆ.

49. HIRUNDO RUSTICA, *Linn.*
Swallow.

50. CHELIDON URBICA, *Linn.*
Martin.

51. COTILE RIPARIA, *Linn.*
Sand Martin.

Family FRINGILLIDÆ.

Sub-family *FRINGILLINÆ.*

52. CARDUELIS ELEGANS, *Steph.*
Goldfinch.

53. CHRYSOMITRIS SPINUS, *Linn.*
Siskin.

54. LIGURINUS CHLORIS, *Linn.*
Greenfinch.

55. COCCOTHRAUSTES VULGARIS, *Pall.*
Hawfinch.

56. PASSER DOMESTICUS, *Linn.*
Common Sparrow.

57. PASSER MONTANUS, *Linn.*
Tree Sparrow.

58. FRINGILLA CŒLEBS, *Linn.*
Chaffinch.

59. FRINGILLA MONTIFRINGILLA, *Linn.*
Brambling.

60. LINOTA CANNABINA, *Linn.*
Linnet.

61. LINOTA RUFESCENS, *Vieill.*
Lesser Redpoll.

62. LINOTA FLAVIROSTRIS, *Linn.*
Twite.

Sub-family *LOXIINÆ.*

63. PYRRHULA EUROPÆA, *Vieill.*
Bullfinch.

64. LOXIA CURVIROSTRA, *Linn.*
Common Crossbill.

Sub-family *EMBERIZINÆ.*

65. EMBERIZA MELANOCEPHALA, *Scop.*
Black-headed Bunting.

66. EMBERIZA MILIARIA, *Linn.*
Common Bunting.

67. EMBERIZA CITRINELLA, *Linn.*
Yellow Bunting.

68. EMBERIZA CIRLUS, *Linn.*
Cirl Bunting.

69. EMBERIZA SCHŒNICLUS, *Linn.*
Reed Bunting.

Family ALAUDIDÆ.

70. ALAUDA ARVENSIS, *Linn.*
Sky Lark.

71. ALAUDA ARBOREA, *Linn.*
Wood-Lark.

Family STURNIDÆ.

72. STURNUS VULGARIS, *Linn.*
Common Starling.

Family CORVIDÆ.

73. PYRRHOCORAX GARRULUS, *Linn.*
Chough.

74. GARRULUS GLANDARIUS, *Linn.*
Jay.

75. PICA RUSTICA, *Scop.*
Magpie.

76. CORVUS MONEDULA, *Linn.*
Jackdaw.

77. CORVUS CORONE, *Linn.*
Carrion Crow.

78. CORVUS FRUGILEGUS, *Linn.*
Rook.

79. CORVUS CORAX, *Linn.*
Raven.

Family CYPSELIDÆ.

80. CYPSELUS APUS, *Linn.*
Swift.

Family CAPRIMULGIDÆ.

81. CAPRIMULGUS EUROPÆUS, *Linn.*
Nightjar.

Family PICIDÆ.

Sub-family *PICINÆ.*

82. PICUS MAJOR, *Linn.*
Great Spotted Woodpecker.

83. PICUS MINOR, *Linn.*
Lesser Spotted Woodpecker.

84. GEOINUS VIRIDIS, *Linn.*
Green Woodpecker.

Sub-family *JYNGINÆ.*

85. JYNX TORQUILLA, *Linn.*
Wryneck.

Family ALCEDINIDÆ.

86. ALCEDO ISPIDA, *Linn.*
Kingfisher.

Family MEROPIDÆ.

87. MEROPS APIASTER, *Linn.*
Bee-eater.

Family UPUPIDÆ.

88. UPUPA EPOPS, *Linn.*
Hoopoe.

Family CUCULIDÆ.

89. CUCULUS CANORUS, *Linn.*
Cuckoo.

Family STRIGIDÆ.

90. STRIX FLAMMEA. *Linn.*
Barn Owl.

Family BUBONIDÆ.

91. ASIO OTUS, *Linn.*
Long-eared Owl.

92. ASIO ACCIPITRINUS, *Pall.*
Short-eared Owl.

93. SYRNIUM ALUCO, *Lin.t.*
Tawny Owl.

Family FALCONIDÆ.

94. CIRCUS ÆRUGINOSUS, *Linn.*
Marsh Harrier.

95. CIRCUS CINERACEUS, *Mont.*
MONTAGU'S HARRIER.

96. CIRCUS CYANEUS, *Linn.*
Hen Harrier.

97. BUTEO VULGARIS, *Leach.*
Common Buzzard.

98. AQUILA CHRYSAETUS, *Linn.*
Golden Eagle.

99. HALIÆTUS ALBICILLA, *Pall.*
Sea Eagle.

100. ACCIPITER NISUS, *Linn.*
Sparrow Hawk.

101. MILVUS ICTINUS, *Sav.*
Kite.

102. PERNIS APIVORUS, *Linn.*
Honey Buzzard.

103. FALCO SUBBUTEO, *Linn.*
Hobby.

104. FALCO ÆSALON, *Tunst.*
Merlin.

105. FALCO TINNUNCULUS, *Linn.*
Kestrel.

106. PANDION HALIÆTUS, *Linn.*
Osprey.

Family PELECANIDÆ.

107. PHALACROCORAX CARBO, *Linn.*
Cormorant.

108. PHALACROCORAX GRACULUS, *Linn.*
Shag.

109. SULA BASSANA, *Linn.*
Gannet.

Family ARDEIDÆ.

110. ARDEA CINEREA, *Linn.*
Heron.

111. BOTAURUS STELLARIS, *Linn.*
Bittern.

Family ANATIDÆ.

112. ANSER CINEREUS, *Meyer.*
Grey-lag Goose.

113. CYGNUS OLOR, *Gmel.*
Mute Swan.

114. TADORNA CORNUTA, *Gmel.*
Sheldrake.

115. ANAS BOSCAS, *Linn.*
Mallard.

116. SPATULA CLYPEATA, *Linn.*
Shoveller.

117. QUERQUEDULA CREOCA, *Linn.*
Common Teal.

118. QUERQUEDULA CIRCIA, *Linn.*
Garganey Teal.

119. MARECA PENELOPE, *Linn.*
Wigeon.

120. FULIGULA FERINA, *Linn.*
Pochard.

121. FULIGULA CRISTATA, *Leach.*
Tufted Duck.

122. CLANGULA GLAUCION, *Linn.*
Golden Eye.

123. SOMATERIA MOLLISSIMA, *Linn.*
Eider Duck.

124. ŒDEMIA NIGRA, *Linn.*
Common Scoter.

125. MERGUS MERGANSER, *Linn.*
Goosander.

126. MERGUS SERRATOR, *Linn.*
Red-breasted Merganser.

Family COLUMBIDÆ.

127. COLUMBA PALUMBUS, *Linn.*
Ring Dove.

128. COLUMBA LIVIA, *Bonn.*
Rock Dove.

129. COLUMBA ÆNAS, *Linn.*
Stock Dove.

130. TURTUR COMMUNIS, *Selby.*
Turtle Dove.

Family PHASIANIDÆ.

131. PHASIANUS COLCHICUS, *Linn.*
Pheasant.

132. CACCABIS RUFA, *Linn.*
Red-legged Partridge.

133. PERDIX CINEREA, *Lath.*
Common Partridge.

134 COTURNIX COMMUNIS, *Bonn.*
Quail.

Family TETRAONIDÆ.

135. LAGOPUS MUTUS, *Leach.*
Ptarmigan.

136. LAGOPUS SCOTICUS, *Lath.*
Red Grouse.

137. TETRAO TETRIX, *Linn.*
Black Grouse.

138. TETRAO UROGALLUS, *Linn.*
Capercaillie.

Family RALLIDÆ.

139. RALLUS AQUATICUS, *Linn.*
Water Rail.

140. PORZANA MARUETTA, *Leach.*
Spotted Crake.

141. PORZANA BAILLONI, *Vieill.*
Baillon's Crake.

142. CREX PRATENSIS, *Bechst.*
Land Rail.

143. GALLINULA CHLOROPUS, *Linn.*
Moorhen.

144. FULICA ATRA, *Linn.*
Coot.

Family OTIDIDÆ.

145. OTIS TARDA, *Linn.*
Great Bustard.

Family ŒDICNEMIDÆ.

146. ŒDICNEMIS SCOLOPAX, *Gmel.*
Stone Curlew.

Family CHARADRIIDÆ.

147. CHARADRIUS PLUVIALIS, *Linn.*
Golden Plover.

148. ÆGIALITIS CANTIANA, *Lath.*
Kentish Plover.

149. ÆGIALITIS HIATICULA, *Linn.*
Ringed Plover.

150. EUDROMIAS MORINELLUS, *Linn.*
Dotterel.

151. VANELLUS VULGARIS, *Bechst.*
Lapwing.

152. STREPSILAS INTERPRES, *Linn.*
Turnstone.

153. HÆMATOPUS OSTRALEGUS, *Linn.*
Oystercatcher.

Family SCOLOPACIDÆ.

154. PHALAROPUS HYPERBOREUS, *Linn.*
Red-necked Phalarope.

155. SCOLOPAX RUSTICOLA, *Linn.*
Woodcock.

156. GALLINAGO CŒLESTIS, *Frenz.*
Snipe.

157. TRINGA ALPINA, *Linn.*
Dunlin.

158. TOTANUS HYPOLEUCUS, *Linn.*
Common Sandpiper.

159. TOTANUS CALIDRIS, *Linn.*
Redshank

160. TOTANUS CANESCENS, *Gmel.*
Greenshank.

161. TOTANUS PUGNAX, *Briss.*
Ruff.

162. LIMOSA ÆGOCEPHALA, *Linn.*
Black-tailed Godwit.

163. NUMENIUS PHÆOPUS, *Linn.*
Whimbrel.

164. NUMENIUS ARQUATA, *Linn.*
Curlew.

Family LARIDÆ.

Sub-family STERNINÆ.

165. STERNA MACRURA, *Naum.*
Arctic Tern.

166. STERNA FLUVIATILIS, *Naum.*
Common Tern.

167. STERNA MINUTA, *Linn.*
Little Tern.

168. STERNA DOUGALLI, *Mont.*
Roseate Tern.

169. STERNA CANTIACA, *Gmel.*
Sandwich Tern.

170. HYDROCHELIDON NIGRA, *Linn.*
Black Tern.

Sub-family LARINÆ.

171. LARUS RIDIBUNDUS, Linn.
Black-headed Gull.

172. LARUS CANUS, Linn.
Common Gull.

173. LABUS ARGENTATUS, Gmel.
Herring Gull.

174. LABUS FUSCUS, Linn.
Lesser Black-backed Gull.

175. LABUS MARINUS, Linn.
Greater Black-backed Gull.

176. RISSA TRIDACTYLA, Linn.
Kittiwake.

Sub-family STERCORARIINÆ.

177. STERCORARIUS CATARRHACTES, Linn.
Common Skua.

178. STERCORARIUS POMATORHINUS, Temm.
Pomatorhine Skua.*

179. STERCORARIUS CREPIDATUS, Banks.
Richardson's Skua.

180. STERCORARIUS PARASITICUS, Linn.
Buffon's Skua.

Family PROCELLARIIDÆ.

181. PROCELLARIA PELAGICA, Linn.
Storm Petrel.

182. PROCELLARIA LEUCORRHOA, Vieill.
Leach's Petrel.

183. PUFFINUS ANGLORUM, Temm.
Manx Shearwater.

184. FULMARUS GLACIALIS, Linn.
Fulmar Petrel.

Family ALCIDÆ.

185. ALCA TORDA, Linn.
Razorbill.

186. LOMVIA TROILE, Linn.
Common Guillemot.

187. UBIA GRYLLE, Linn.
Black Guillemot.

188. FRATERCULA ARCTICA, Linn.
Puffin.

Family COLYMBIDÆ.

189. COLYMBUS GLACIALIS, Linn.
Great Northern Diver.

190. COLYMBUS ARCTICUS, Linn.
Black-throated Diver.

191. COLYMBUS SEPTENTRIONALIS, Linn.
Red-throated Diver.

Family PODICIPITIDÆ.

192. PODICEPS CRISTATUS, Linn.
Great Crested Grebe.

193. PODICEPS FLUVIATILIS, Tunst.
Little Grebe.

* This should probably be omitted as a species breeding in the British Isles.

INDEX TO ENGLISH NAMES.

Q

WEST, NEWMAN AND CO., PRINTERS, HATTON GARDEN, LONDON, E.C.